Introductory Chemistry in the Laboratory

Steven S. Zumdahl
University of Illinois Urbana-Champaign

Donald J. DeCoste
University of Illinois Urbana-Champaign

Written by

James F. Hall
University of Massachusetts Lowell

BROOKS/COLE
CENGAGE Learning

Australia • Brazil • Japan • Korea • Mexico • Singapore • Spain • United Kingdom • United States

ISBN-13: 978-0-538-73642-8
ISBN-10: 0-538-73642-9

Brooks/Cole
20 Channel Center Street
Boston, MA 02210
USA

Cengage Learning is a leading provider of customized learning solutions with office locations around the globe, including Singapore, the United Kingdom, Australia, Mexico, Brazil, and Japan. Locate your local office at:
www.cengage.com/global

Cengage Learning products are represented in Canada by Nelson Education, Ltd.

To learn more about Brooks/Cole, visit
www.cengage.com/brookscole

Purchase any of our products at your local college store or at our preferred online store
www.ichapters.com

Printed in the United States of America
3 4 5 6 7 8 9 16 15 14 13 12

Contents

Contents

Preface

Introduction: Why We Study Chemistry

On the simplest level, we say that chemistry is the study of matter and the transformations it undergoes. On a more intellectual level, we realize that chemistry can explain (or try to explain) such things as the myriad chemical reactions in the living cell, the transmission of energy by superconductors, the working of transistors, and even how the oven and drain cleaners we use in our homes function.

The study of chemistry is required for students in many other fields because it is a major unifying force among these other subjects. Chemistry is the study of matter itself. In other disciplines, particular *aspects* of matter or its applications are studied, but the basis for such study rests in a firm foundation in chemistry.

Laboratory study is required in chemistry courses for several reasons. In some cases, laboratory work may serve as an introduction to further, more difficult laboratory work (this is true of the first few experiments in this laboratory manual). Before you can perform meaningful or relevant experiments, you must learn to use the basic tools of the trade.

Sometimes laboratory experiments in chemistry are used to demonstrate the topics covered in course lectures. For example, every chemistry course includes a lengthy exposition on the properties of gases and a discussion of the gas laws. A laboratory demonstration of these laws may help to clarify them.

Laboratory work can teach you the various standard techniques used by scientists in chemistry and in most other fields of science. For example, pipets are used in many biological, health and engineering disciplines when a precisely measured volume of liquid is needed. Chemistry lab is an excellent place to learn techniques correctly; you can learn by practicing techniques in general chemistry lab rather than on the job later.

Finally, your instructor may use the laboratory as a means of judging how much you have learned and understood from your lectures. Sometimes students can read and understand their textbook and lecture notes and do reasonably well on examinations without gaining much practical knowledge of the subject. The lab serves as a place where classroom knowledge can be synthesized and applied to realistic situations.

What Will Chemistry Lab Be Like?

Undergraduate chemistry laboratories are generally 2 to 3 hours long, and most commonly meet once each week. The laboratory is generally conducted by a teaching assistant – a graduate student with several years of experience in chemistry laboratory. Smaller colleges, which may not have a graduate program in chemistry, may employ instructors whose specific duties consist of running the general chemistry labs, or upperclass student assistants who work under the course professor's supervision. Your professor undoubtedly will visit the laboratory frequently to make sure that everything is running smoothly.

The first meeting of a laboratory section is often quite hectic. This experience is perfectly normal, so do not judge your laboratory by this first meeting. Part of the first lab meeting will be devoted to checking in to the laboratory. You will be assigned a workspace and locker and will be asked to go through the locker to make sure that it contains all the equipment you will need. Your instructor will then conduct a brief orientation to the laboratory, giving you the specifics of how the laboratory period will be operated. Be sure to ask any questions you may have about the operation of the laboratory.

Your instructor will discuss with you the importance of safety in the laboratory and will point out the emergency equipment. *Pay close attention to this discussion so that you will be ready for any eventuality.*

Finally, your instructor will discuss the experiment you will be performing after check-in. Although we have tried to make this laboratory manual as explicit and complete as possible, your instructor will certainly offer tips or advice based on his or her own experience and expertise. Take advantage of this information.

How Should You Prepare for Lab?

Laboratory should be one of the most enjoyable parts of your study of chemistry. Instead of listening passively to your instructor, you have a chance to witness chemistry in action. However, if you have not suitably prepared for lab, you will spend most of your time wondering what is going on.

This manual has been written and revised to help you prepare for laboratory. First of all, each experiment has a clearly stated *Objective*. This section is a very short summary of the lab's purpose. This *Objective* will suggest what sections of your textbook or lecture notes might be appropriate for review before lab. Each experiment also contains an *Introduction* that reviews explicitly some of the theory and methods to be used in the experiment. You should read through this *Introduction* before the laboratory period and make cross-references to your textbook while reading. Make certain that you understand fully what an experiment is supposed to demonstrate before you attempt that experiment.

The lab manual contains a detailed *Procedure* for each experiment. The *Procedures* have been written to be as clear as possible, but do not wait until you are actually performing the experiment to read this material. Study the *Procedure* before the lab period, and question the instructor before the lab period on anything you do not understand. Sometimes it is helpful to write up a summary or overview of a long procedure. This preparation can prevent major errors when you are actually performing the experiment. Another technique that students use in getting ready for lab is to prepare a flow chart, indicating the major procedural steps, and any potential pitfalls, in the experiment. You can keep the flow chart handy while performing the experiment to help avoid any gross errors. A flow chart is especially useful for planning your time most effectively: for example, if a procedure calls for heating something for an hour, a flow chart can help you plan what else you can get done during that hour.

This lab manual contains a *Laboratory Report* for each of the experiments. Each *Laboratory Report* is preceded by a set of *Pre-Laboratory Questions*, which your instructor will probably ask you to complete before coming to lab. Sometimes these questions will include numerical problems similar to those you will encounter in processing the data you will be collecting in the experiment. Obviously, if you review a calculation before lab, things will be that much easier when you are actually in the lab.

I hope you will find your study of chemistry to be both meaningful and enjoyable. As always, I encourage those who use this manual – both instructors and students – to assist me in improving it for the future. Any comments or criticism will be greatly appreciated.

James F. Hall

Laboratory Glassware and Other Apparatus

Introduction

When you open your laboratory locker for the first time, you are likely to be confronted with a bewildering array of various sizes and shapes of glassware and other apparatus. Glass is used more than any other material for the manufacture of laboratory apparatus because it is relatively cheap, usually easy to clean, and can be heated to fairly high temperatures or cooled to quite low temperatures. Most importantly, glass is used because it is impervious to and non-reactive with most reagents encountered in the beginning chemistry laboratory. Most glassware used in chemistry laboratories is made of borosilicate glass, which is relatively sturdy and safe to use at most temperatures. Such glass is sold under such trade names as Pyrex® (Corning Glass Co.) and Kimax® (Kimble Glass Co.). If any of the glassware in your locker does not have either of these trade names marked on it, consult with your instructor before using it at temperatures above room temperature.

Many pieces of laboratory glassware and apparatus have special names that have evolved over the centuries as chemistry has been studied. You should learn the names of the most common pieces of laboratory apparatus to make certain that you use the correct equipment for the experiments in this manual. Study the drawings of common pieces of laboratory apparatus that are shown on the next few pages. Compare the equipment in your locker with these drawings, and identify all the pieces of apparatus that have been provided to you. It might be helpful for you to label the equipment in your locker, at least for the first few weeks of the term.

While your locker may not contain all the apparatus in the drawings, be sure to ask your instructor for help if there is equipment in your locker that is not described in the drawings. While examining your locker equipment, watch for chipped, cracked or otherwise imperfect glassware. *Imperfect glassware is a major safety hazard, and must be discarded.* Don't think that because a beaker has just a little crack, it can still be used. *Replace all glassware that has any cracks, chips, star fractures, or any other deformity.* Most college and university laboratories will replace imperfect glassware free of charge during the first laboratory meeting of the term, but may assess charges if breakage is discovered later in the term.

You may wish to clean the set of glassware in your locker during the first meeting of your laboratory section, but be warned that lab glassware has an annoying habit of becoming dirty again without apparent human intervention. Glassware always looks clean when wet, but tends to dry with water spots (and show minor imperfections in the glass that are not visible when wet). It is best to thoroughly clean all the locker glassware at the start, and then rinse the glassware out before its first use. In the future, clean glassware before leaving lab for the day, and rinse before using. Instructions for the proper cleaning of laboratory glassware are provided after the diagrams of apparatus.

Laboratory Glassware/Apparatus

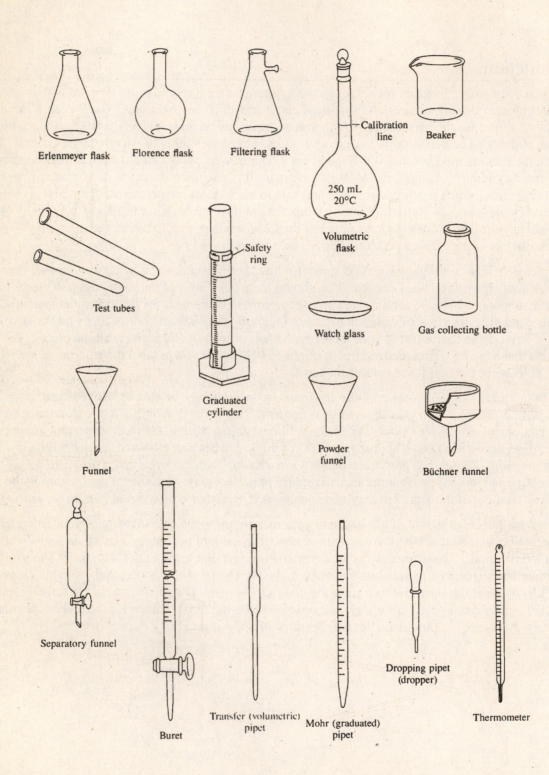

Erlenmeyer flask

Florence flask

Filtering flask

Volumetric flask

Calibration line

250 mL 20°C

Beaker

Test tubes

Graduated cylinder

Safety ring

Watch glass

Gas collecting bottle

Funnel

Powder funnel

Büchner funnel

Separatory funnel

Buret

Transfer (volumetric) pipet

Mohr (graduated) pipet

Dropping pipet (dropper)

Thermometer

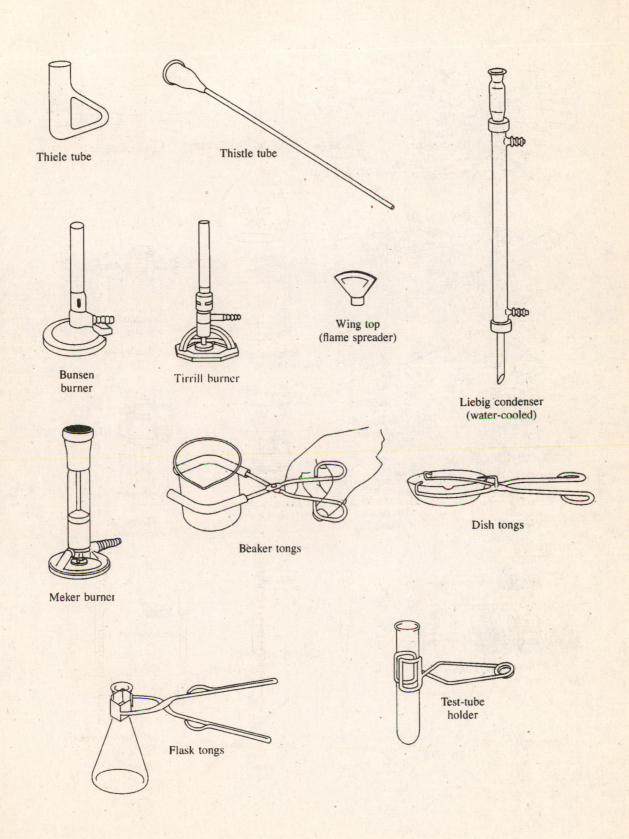

Thiele tube

Thistle tube

Bunsen burner

Tirrill burner

Wing top (flame spreader)

Liebig condenser (water-cooled)

Meker burner

Beaker tongs

Dish tongs

Flask tongs

Test-tube holder

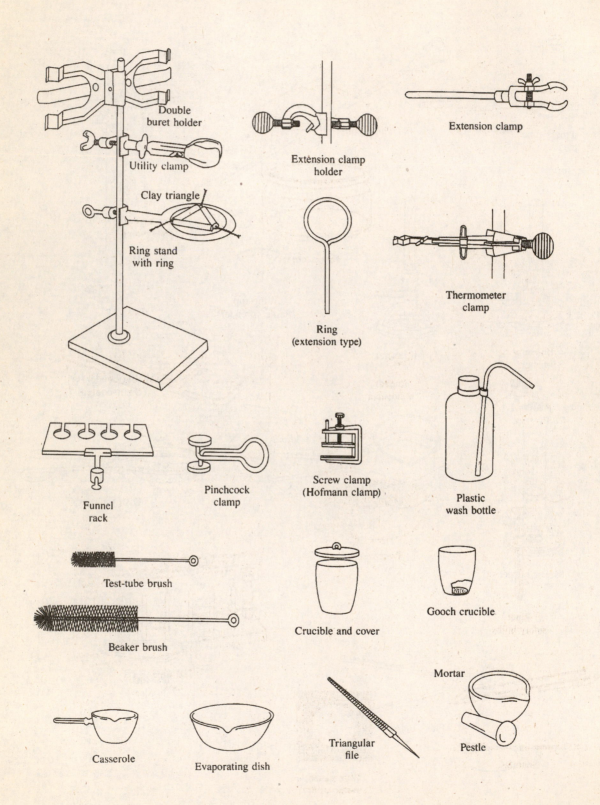

Double buret holder

Utility clamp

Clay triangle

Ring stand with ring

Extension clamp holder

Extension clamp

Ring (extension type)

Thermometer clamp

Funnel rack

Pinchcock clamp

Screw clamp (Hofmann clamp)

Plastic wash bottle

Test-tube brush

Beaker brush

Crucible and cover

Gooch crucible

Casserole

Evaporating dish

Triangular file

Mortar

Pestle

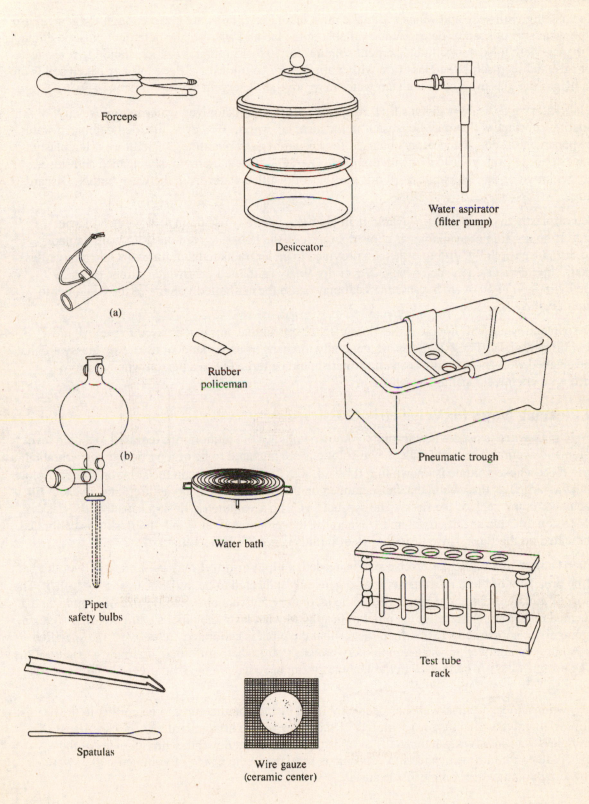

Forceps

Desiccator

Water aspirator
(filter pump)

(a)

Rubber
policeman

Pneumatic trough

(b)

Water bath

Pipet
safety bulbs

Test tube
rack

Spatulas

Wire gauze
(ceramic center)

Cleaning of Laboratory Glassware

General Information

A simple washing with soap and water will make most laboratory glassware clean enough for general use. Determine what sort of soap or detergent is available in the lab and whether the detergent must be diluted. Allow the glassware to soak in dilute detergent solution for 10 to 15 minutes (which should remove any grease or oil), and then scrub the glassware with a brush from your locker if there are any caked solids on the glass. Rinse the glassware well with tap water to remove all detergent.

Usually laboratory glassware is given a final rinse with distilled or deionized water to remove any contaminating substances that may be present in the local tap water. However, distilled/deionized water is very expensive to produce and cannot be used for rinsing in great quantities. Therefore, fill a plastic squeeze wash bottle from your locker with distilled water, and rinse the previously cleaned and rinsed glassware with two or three separate 5–10 mL portions of distilled water from the wash bottle. Discard the rinsings.

If wooden or plastic drying hooks are available in your lab, you may use them to dry much of your glassware. If hooks are not available, or items do not fit on the hooks, spread out glassware on paper towels to dry. Occasionally, drying ovens are provided in undergraduate laboratories. Only simple flasks and beakers should be dried in such ovens. Never dry finely calibrated glassware (burets, pipets, graduated cylinders) in an oven, because the heat may cause the calibrated volume of the container to change appreciably.

If simple soap and water will not clean the glassware, chemical reagents can be used to remove most stains or solid materials. These reagents are generally too dangerous for student use. Any glassware that cannot be cleaned well enough with soap and water should be turned in to a person who has more experience with chemical cleaning agents.

Special Cleaning Notes for Volumetric Glassware

Volumetric glassware is used when absolutely known volumes of solutions are required to a high level of precision and accuracy. For example, when solutions are prepared to be of a particular concentration, a volumetric flask whose volume is known to ± 0.01 mL may be used to contain the solution. The absolute volume that a particular flask will contain is stamped on the flask by the manufacturer, and an exact **fill mark** is etched on the neck of the flask. The symbol "TC" on a volumetric flask means that the flask is intended "to contain" the specified volume. Generally, the temperature at which the flask was calibrated is also indicated on the flask, since the volumes of liquids vary with temperature.

If a solution sample of a particular precise size is needed, a **pipet** or **buret** may be used to deliver the sample (the precision of these instruments is also generally indicated to be to the nearest ± 0.01 mL). The normal sort of transfer volumetric pipet used in the laboratory delivers one specific size of sample, and a mark is etched on the upper barrel of the pipet indicating to what level the pipet should be filled. Such a pipet is generally also marked "TD," which means that the pipet is calibrated "to deliver" the specified volume. A buret can deliver any size sample very precisely, from zero milliliters up to the capacity of the buret. The normal Class A buret used in the laboratory can have its volume read precisely to the nearest 0.01 mL.

Obviously, when a piece of glassware has been calibrated by the manufacturer to be correct to the nearest 0.01 milliliter, the glassware must be absolutely clean before use. The standard test for cleanliness of volumetric glassware involves watching a film of distilled water run down the interior sides of the glassware. Water should flow in sheets (a continuous film) down the inside of volumetric glassware, without beading up anywhere on the inside surface.

If water beads up anywhere on the interior of volumetric glassware, the glassware must be soaked in dilute detergent solution, scrubbed with a brush (except for pipets), and rinsed thoroughly (both with tap water and distilled water). The process must be repeated until the glassware is absolutely clean. Narrow bristled brushes are available for reaching into the long, narrow necks of volumetric flasks, and special long-handled brushes are available for scrubbing burets. Since brushes cannot be fitted into the barrel of pipets, if it is not possible to clean the pipet completely on two or three attempts, the pipet should be exchanged for a new one (special pipet washers are probably available in the stockroom for cleaning pipets that have been turned in by students as uncleanable).

Volumetric glassware is generally used wet. Rather than drying the glassware (possibly allowing the glassware to become water-spotted, which may cause incorrect volumes during use), the user rinses the glassware with whatever is going to be used ultimately in the glassware. For example, if a buret will be filled with standard acid solution, the buret – still wet from cleaning – is rinsed with small portions of the same acid (the rinsings are discarded). Rinsing with the solution that is going to be used with the glassware insures that no excess water will dilute the solution to be measured. *Volumetric glassware is never heated in an oven,* since the heat may destroy the integrity of the glassware's calibration. If space in your locker permits, you might wish to leave volumetric glassware, especially burets, filled with distilled water between laboratory periods. A buret that has been left filled with water will require much less time to clean for subsequent use (consult with the instructor to see if this is possible).

Safety in the Chemistry Laboratory

Introduction

Chemistry is an experimental science. You cannot learn chemistry without getting your hands dirty. Beginning chemistry students face the prospect of laboratory work with some apprehension. It would be untruthful to say that there is no element of risk in a chemistry lab. Chemicals can be dangerous. The more you study chemistry, the larger the risks will become. But if you approach your laboratory work calmly and studiously, you will minimize the risks.

During the first laboratory meeting, you should ask your instructor for a brief tour of the laboratory room. Ask him or her to point out for you the locations of the various pieces of emergency apparatus provided by your college or university. At your bench, construct a map of the laboratory, noting the location of the exits from the laboratory and the location of all safety equipment. Close your eyes, and test whether you can locate the exits and safety equipment from memory. A brief discussion of the major safety apparatus and safety procedures follows. A Safety Quiz is provided at the end of this section to test your comprehension and appreciation of this material. For additional information on safety in your particular laboratory, consult with your laboratory instructor or course professor.

Protection for the Eyes

Government regulations, as well as common sense, demand the wearing of protective eyewear while you are in the laboratory. Such eyewear must be worn even if you personally are not working on an experiment. Figure 1 shows one common form of plastic **safety goggle.**

Figure 1. A typical student plastic safety goggle.

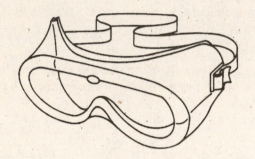

Although you may not use the particular type of goggle shown in the figure, your eyewear must include shatterproof lenses and side shields that will protect you from splashes. *Safety glasses must be worn at all times while you are in the laboratory, whether or not you are working with chemicals.* Failure to wear safety glasses may result in your being failed or withdrawn from your chemistry course, or in some other disciplinary action.

In addition to protective goggles, an **eyewash fountain** provides eye protection in the laboratory. If a chemical splashes near your eyes, you should use the eyewash fountain before the material has a chance to reach your eyes. If the eyewash fountain is not near your bench, wash your eyes quickly but thoroughly with water from the nearest source, and then use the eyewash. Figure 2 shows an eyewash fountain.

Figure 2. Laboratory emergency eyewash fountain

The eyewash has a panic bar that enables the eyewash to be activated easily in an emergency. If you need the eyewash, don't be modest – *use it immediately*. It is critical that you protect your eyes quickly.

Protection from Fire

The danger of uncontrolled fire in the chemistry laboratory is very real, since the lab typically has a fairly large number of flammable liquids in it, and open-flame gas burners are generally used for heating (see later in this discussion for proper use of the gas burner). With careful attention, however, the danger of fire can be reduced considerably. Always check around the lab before lighting a gas burner to ensure that no one is pouring or using any flammable liquids. Be especially aware that the vapors of most flammable liquids are heavier than air and tend to concentrate in sinks (where they may have been poured) and at floor level. Since your laboratory may be used by other classes, always check with your instructor before beginning to use gas burners.

Despite precautions being taken, there may be a fire. The method used to fight fires depends on the size of the fire and on the substance that is burning. If only a few drops of flammable liquid have been accidentally ignited, and no other reservoir of flammable liquid is nearby, the fire can usually be put out by covering it with a beaker. This action deprives the fire of oxygen and will usually extinguish the fire in a few minutes. Leave the beaker in place for several minutes to ensure that the flammable material has cooled and will not flare up again.

In the unlikely event that a larger chemical fire occurs, carbon dioxide **fire extinguishers** are available in the lab (usually mounted near one of the exits from the room). An example of a typical carbon dioxide fire extinguisher is shown in Figure 3.

Figure 3. A typical carbon dioxide fire extinguisher
Pull the metal ring to release the extinguisher handle)

Before activating the extinguisher, pull the metal safety ring from the handle. Direct the extinguisher toward the base of the flames. The carbon dioxide not only smothers the flames, it also cools the flammable material quickly. If it becomes necessary to use the fire extinguisher, be sure to turn the extinguisher in at the stockroom so that it can be refilled immediately afterward. If the carbon dioxide extinguisher does not immediately put out the fire, evacuate the laboratory and call the fire department.

Carbon dioxide fire extinguishers must *not* be used on fires involving magnesium or certain other reactive metals, since carbon dioxide may react vigorously with the burning metal and make the fire worse.

One of the most frightening and potentially tragic fire-related accidents is the igniting of a person's clothing. For this reason, certain types of clothing are *not appropriate* for the laboratory and must not be worn. Since sleeves are most likely to come in closest proximity to flames, any garment that has bulky or loose sleeves should not be worn in the laboratory. Certain fabrics should also be avoided; silk and certain synthetic materials may be highly flammable. Ideally, students should be asked to wear laboratory coats with tightly fitting sleeves made specifically for the chemistry laboratory. This clothing may be required by your particular college or university. Long hair also presents a clear danger if it is allowed to hang loosely in the vicinity of the flame. Long hair must be pinned back or held with a rubber band.

In the unlikely event a student's clothing or hair is ignited, his or her neighbors must take prompt action to prevent severe burns. Most laboratories have two options for extinguishing such fires: the **water shower** and the **fire blanket**. Figure 4 shows a typical laboratory emergency water shower:

Figure 4. Laboratory emergency deluge shower
Use the shower to extinguish clothing fires and in the event of a large-scale chemical spill.

Showers such as this generally are mounted near the exits from the laboratory. In the event that someone's clothing or hair is on fire, immediately push or drag the person to the shower and pull the metal ring of the shower. Safety showers generally dump 40 to 50 gallons of water, which should extinguish the flames. Be aware that the showers cannot be shut off once the metal ring has been pulled. For this reason, the shower cannot be demonstrated. (But note that the showers are checked for correct operation on a regular basis.)

Figure 5 shows the other possible apparatus for extinguishing clothing fires – the **fire blanket**. The fire blanket is intended to smother flames. Since it must be operated by the person on fire (by pulling the handle of the fire blanket and wrapping it around oneself), it is not the preferred method of dealing with such an event.

Figure 5. Fire blanket
The blanket is wrapped around the body to smother flames.

Protection from Chemical Burns

Most acids, alkalis and oxidizing and reducing agents must be assumed to be corrosive to skin. It is impossible to avoid these substances completely, since many of them form the backbone of the study of chemistry. Usually, a material's corrosiveness is proportional to its concentration. Most of the experiments in this manual have been set up to use as dilute a mixture of reagents as possible, but this precaution does not entirely remove the problem. Make it a personal rule to *wash your hands regularly* after using any chemical substance and to wash immediately, with plenty of water, if any chemical substance is spilled on the skin.

After working with a substance known to be particularly corrosive, you should wash your hands immediately, even if you did not spill the substance. Someone else using the bottle of reagent may have spilled the substance on the side of the bottle. It is also good practice to hold bottles of corrosive substances with a paper towel or to wear plastic gloves during pouring. Do not make the mistake of thinking that dilute acid cannot burn your skin. Some of the acids used in the laboratory are not volatile, and as water evaporates from a spill, the acid becomes concentrated enough to damage skin. Whenever a corrosive substance is spilled on the skin, you should inform the instructor immediately. If there is any sign of damage to the skin, you will be sent to your college's health services for evaluation by a physician.

In the event of a major chemical spill in which substantial portions of the body or clothing are affected, you must use the emergency water shower. Forget about modesty, and get under the shower immediately.

Protection from Toxic Fumes

Many volatile substances have toxic vapors. A rule of thumb for the chemistry lab is, "If you can smell it, it can probably hurt you." Some toxic fumes (such as ammonia) can overpower you immediately, whereas other toxic fumes are more insidious. A substance does not need to have a bad odor to do severe damage to the respiratory system. There is absolutely no need to expose yourself to toxic fumes. All chemistry laboratories are equipped with **fume exhaust hoods**. A typical hood is indicated in Figure 6.

Figure 6. A common type of laboratory fume exhaust hood
Use the hood whenever a reaction involves toxic fumes. Keep the glass window of the hood partially closed to provide for rapid flow of air.

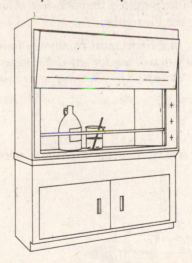

The exhaust hood has fans that draw vapors out of the hood and away from the user. The hood is also used when flammable solvents are required for a given procedure, since the hood will remove the vapors of such solvents from the laboratory and reduce the hazard of fire. The hood is equipped with a safety-glass window that can be used as a shield from reactions, should they become too vigorous. Naturally, the number of exhaust hoods available in a particular laboratory is limited, but *never neglect to use the hood if it is called for*, merely to save a few minutes of waiting time. Finally, reagents are sometimes stored in a hood, especially if the reagents emit toxic fumes. Be sure to return such reagents to the designated hood after use.

Protection from Cuts and Simple Burns

Perhaps the most common injuries to students in the beginning chemistry laboratory are simple cuts and burns. Glass tubing and glass thermometers are used in nearly every experiment and are often not prepared or used properly. Most glass cuts occur when the glass (or thermometer) is being inserted through rubber stoppers in the construction of an apparatus. Use glycerine as a lubricant when inserting glass through rubber (several bottles of glycerine will be provided in your lab). Glycerine is a natural product of human and animal metabolism; it may be applied liberally to any piece of glass. Glycerine is water soluble, and though it is somewhat messy, it washes off easily. You should always remove glycerine before using an apparatus since the glycerine may react with the reagents being used.

Most simple burns in the laboratory occur when a student forgets that an apparatus is hot and touches it. Never touch an apparatus that has been heated until it has cooled for at least five minutes, or unless specific tongs for the apparatus are available.

Report any cuts or burns, no matter how seemingly minor, to the instructor immediately. If there is any visible damage to the skin, you will be sent to your college's health services for immediate evaluation by a physician. What may seem like a scratch may be adversely affected by chemical reagents or may become infected; it must be attended to by trained personnel.

Proper Use of the Laboratory Burner

The laboratory burner is one of the most commonly used pieces of apparatus in the general chemistry laboratory, and may pose a major hazard if not used correctly and efficiently.

The typical laboratory burner is correctly called a **Tirrel burner** (though the term Bunsen burner is often used in a generic sense). A representation of a Tirrel burner is indicated in Figure 7. Compare the burner you will be using with the burner shown in the figure, and consult with your instructor if there seems to be any difference in construction. Burners from different manufacturers may differ slightly in appearance and operation from the one shown in the illustration.

Figure 7. A Tirrel burner of the sort most commonly found in student laboratories
Compare this burner with the one you will use for any differences in construction or operation. The temperature of the flame is increased by allowing air to mix with the gas being burned.

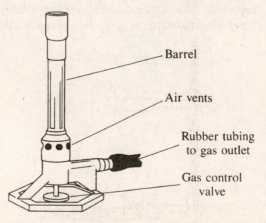

Barrel

Air vents

Rubber tubing
to gas outlet

Gas control
valve

Most laboratories are supplied with natural gas, which consists primarily of the hydrocarbon methane (CH_4). If your college or university is some distance from natural gas lines, your laboratory may be equipped with bottled liquefied gas, which consists mostly of the hydrocarbon propane (C_3H_8). In this case, your burner may have modifications to allow the efficient burning of propane.

A length of thin-walled rubber tubing should attach your burner to the gas main jet. If your burner has a screw-valve on the bottom for controlling the flow of gas, the valve should be closed (by turning in a right-hand direction) before you light the burner. The barrel of the burner should be rotated to close the air vents (or closed by sliding the air vent cover over the vent holes, if your burner has this construction).

To light the burner, turn the gas main jet to the open position. If your burner has a screw-valve on the bottom, open this valve until you hear the hiss of gas. Without delay, use your striker or a match to light the burner. If the burner does not light on the first attempt, shut off the gas main jet and make sure that all rubber tubing connections are tight. Then reattempt to light the burner. You may have to increase the flow of gas using the screw-valve on the bottom of the burner.

After lighting the burner, the flame is likely to be yellow and very unstable (easily blown about by drafts in the lab). The flame at this point is relatively cool (the gas is barely burning) and would be very inefficient in heating. To make the flame hotter and more stable, open the air vents on the barrel of the burner slowly to allow oxygen to mix with the gas before combustion, which should cause the flame's size to decrease and its color to change. A proper mixture of air and gas gives a pale blue flame, with a bright blue, cone-shaped region in the center. The hottest part of the flame is directly above the tip of the bright blue cone. Whenever an item is to be heated strongly, it should be positioned directly above this bright blue cone. You should practice adjusting the flame to get the ideal pale blue flame with its blue inner cone, using the control valve on the bottom of the burner or the gas main jet.

Protection from Apparatus Accidents

An improperly constructed apparatus can create a major hazard in the laboratory, not only to the person using the apparatus, but to his or her neighbors as well. Any apparatus you set up should be constructed exactly as indicated in this manual. If you have any question as to whether you have set up an apparatus correctly, ask your instructor to check the apparatus before you begin the experiment.

Perhaps the most common apparatus accident in the lab is the tipping over of a flask or beaker while it is being heated or otherwise manipulated. All flasks should be *clamped securely* with an adjustable clamp to a ring support. A securely clamped flask is indicated in Figure 8. Be aware that a vacuum flask is almost guaranteed to tip over during suction filtration if it is not clamped, resulting in loss of the crystals being filtered and requiring that you begin the experiment again.

Figure 8. One method of supporting a flask
Be sure to clamp all glassware securely to a ringstand before using.

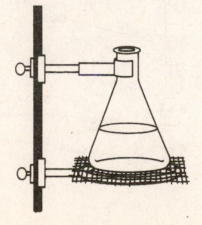

An all-too-common lab accident occurs when a liquid must be heated to boiling in a test tube. Test tubes hold only a few milliliters of liquid and require only a few seconds of heating to reach the boiling point. A test tube cannot be heated strongly in the direct full heat of the burner flame. The contents of the test tube will super-heat and blow out of the test tube like a bullet from a gun. Ideally, when a test tube requires heating, a **boiling water bath** in a beaker should be used. If this setup is not feasible, hold the test tube at a 45° angle a few inches above the flame and heat briefly, keeping the test tube moving constantly (from top to bottom, and from side to side) through the flame during the heating. *Aim the mouth of the test tube away from yourself and your neighbors.* See Figure 9.

Figure 9. Method for hearing a test tube containing a small quantity of liquid
Heat only for a few seconds, beginning at the surface of the liquid and moving downwards. Keep the test tube moving through the flame. Aim the mouth of the test tube away from yourself and your neighbors.

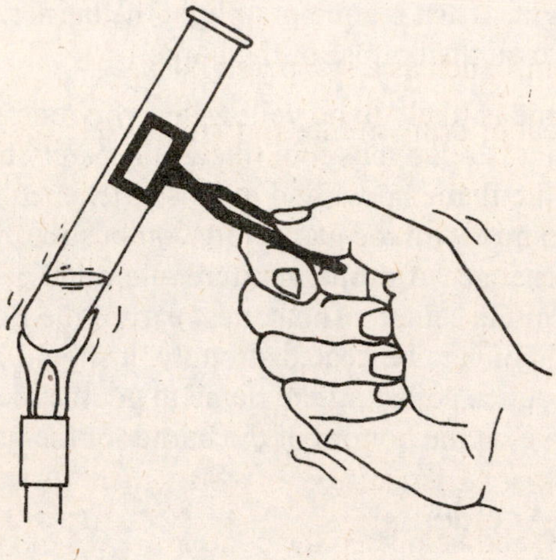

Safety Regulations

- Wear safety glasses at all times while in the laboratory.

- Do not wear short skirts, shorts or bare-midriff shirts in the laboratory.

- Do not wear scarves or neckties in the lab, as they may be ignited accidentally in the burner flame.

- Men who have long beards must secure them away from the burner flame.

- Open-toed shoes and sandals, as well as thin canvas sneakers, are not permitted in the laboratory.

- Never leave Bunsen burners unattended when lighted.

- Never heat solutions to dryness except in an evaporating dish on a hotplate or over a boiling water bath.

- Never heat a "closed system," such as a stoppered flask.

- Never smoke, chew gum, eat or drink in the laboratory; you may inadvertently ingest some chemical substance.

- Always use the smallest amount of substance required for an experiment; more is never better in chemistry. Never return unused portions of a reagent to their original bottle.

- Never store chemicals in your laboratory locker unless you are specifically directed to do so by the instructor.

- Never remove any chemical substance from the laboratory. In many colleges, removal of chemicals from the laboratory is grounds for expulsion or other severe disciplinary action.

- Keep your work area clean, and help keep the common areas of the laboratory clean. If you spill something in a common area, remember that this substance could injure someone else.

- Never fully inhale the vapors of any substance. Waft a tiny amount of vapor toward your nose.

- To heat liquids, always add two or three boiling stones to make the boiling action smoother.

- Never add water to a concentrated reagent when diluting the reagent. Always add the reagent to the water. If water is added to a concentrated reagent, local heating and density effects may cause the water to be splashed back.

- Never work in the laboratory unless the instructor is present. If no instructor is present during your assigned work time, report this to the senior faculty member in charge of your course.

- Never perform any experiment that is not specifically authorized by your instructor. Do not play games with chemicals!

- Dispose of all reaction products as directed by the instructor. In particular, observe the special disposal techniques necessary for flammable or toxic substances.

- Dispose of all glass products in the special container provided.

Information on Hazardous Substances and Procedures

Many of the pre-laboratory assignments in this manual require that you use one of the many chemical handbooks available in any scientific library to look up various data for the substances you will be working with. You should also use such handbooks as a source of information about the hazards associated with the substances and procedures you will be using. A particularly useful general reference is the booklet *Safety in Academic Laboratories*, published by the American Chemical Society (your course professor or laboratory instructor may have copies).

There are also many useful online sources available. For example, Material Safety Data Sheets (MSDS) for many chemicals are available online from the various chemical supply companies. Many chemistry journals, textbooks and handbooks are also available online. Check the Web site of your school's library: many libraries have subscriptions to online databases that you can access.

Each of the experiments in this manual includes a section entitled *Safety Precautions,* which gives important information about expected hazards. This manual also lists some of the hazards associated with common chemical substances in Appendix J. Problems such as flammability and toxicity are described in terms of a low/medium/high rating scale. If you have any questions about this material, consult with your instructor before the laboratory period. Safety is *your* responsibility.

You may now be even more hesitant about beginning your chemistry laboratory experience, now that you have read all these warnings. Be cautious…be careful…be thoughtful…but don't be afraid. Every precaution will be taken by your instructor and your college or university. Realize, however, that *you* bear the ultimate responsibility for your own safety in the laboratory.

Name: _____ Section: _____

Lab Instructor: _____ Date: _____

Safety Quiz

1. Sketch below a representation of your laboratory, showing clearly the location of the following: exits, fire extinguishers, safety shower, eyewash fountain and fume hoods.

2. Why must you wear safety glasses or goggles at all times while you are in the laboratory, even when you are not personally working on an experiment?

    _____

    _____

    _____

    _____

3. Suppose your neighbor in the lab were to spill several milliliters of a flammable substance near her burner, and the substance ignited. How should this relatively small fire be extinguished?

    _____

    _____

    _____

4. What steps should be taken if a larger scale fire were to break out in the laboratory?

    _____

    _____

    _____

    _____

5. What measures should be taken if a student's clothing ignited?

    _____
    _____
    _____
    _____

6. List five types of clothing or footwear that are *not* acceptable in the laboratory, and explain why they are not acceptable.

    _____
    _____
    _____
    _____
    _____
    _____
    _____
    _____
    _____

7. Suppose you were pouring concentrated nitric acid from a bottle into your reaction flask, and you spilled the acid down the front of your shirt. What should you do?

    _____
    _____
    _____

8. What steps should you have taken if you had spilled the nitric acid on your hand rather than on your clothing?

    _____
    _____
    _____
    _____
    _____

9. What are the most common student injuries in the chemistry lab, and how can they be prevented?

_____

_____

_____

_____

_____

10. Where should reactions involving the evolution of toxic gases be performed? Why?

_____

_____

_____

_____

11. Why should you immediately clean up any chemical spills in the laboratory?

_____

_____

_____

_____

_____

12. Why should you never deviate from the published procedure for an experiment?

_____

_____

_____

_____

_____

13. Why should apparatus always be clamped securely to a metal ring stand before starting a chemical reaction?

_____

_____

_____

_____

_____

14.   Why should you never eat, drink or smoke while you are in the laboratory?

_____

_____

_____

_____

_____

_____

_____

15.   Suppose the experiment you are to perform in a given lab period involves substances that are toxic or corrosive. What should you do *before* the laboratory to prepare yourself for using these substances safely?

_____

_____

_____

_____

_____

_____

_____

_____

_____

EXPERIMENT 1

The Laboratory Balance: Mass Determinations

Objective

The measurement of mass is fundamentally important in any chemistry laboratory. In this experiment, you will learn how to use the balances in your particular lab.

Introduction

The accurate determination of mass is one of the most fundamental techniques for students of experimental chemistry. **Mass** is a direct measure of the amount of matter in a sample of substance; the mass of a sample is a direct indication of the number of atoms or molecules the sample contains. Since chemical reactions occur in proportion to the number of atoms or molecules of reactant present, it is essential that the mass of reactant used in a process be accurately known.

There are various types of balances available in the typical general chemistry laboratory. Such balances differ in their construction, appearance, operation and in the level of precision they permit in mass determinations. Three of the most common types of laboratory balance are indicated in Figures 1-1, 1-2 and 1-3. Determine which sort of balance your laboratory is equipped with, and ask your instructor for a demonstration of the use of the balance if you are not familiar with its operation.

There are some general points to keep in mind when using any laboratory balance:

1. Always make sure that the balance gives a reading of 0.000 grams when nothing is on the balance pan. Adjust the tare or zero knob if necessary. If the balance cannot be set to read zero, ask the instructor for help.

2. All balances, but especially electrical/electronic balances, are damaged by moisture. Do not pour liquids in the immediate vicinity of the balance. Clean up any spills immediately from the balance area. Unless the sample is fully contained in a spill-proof container, liquids should not be weighed on an electronic balance.

3. No reagent chemical substance should ever be weighed directly on the pan of the balance. Ideally, reagents should be weighed directly into the beaker or flask in which they are to be used. Plastic weighing boats may also be used if several reagents are required for an experiment. Pieces of filter paper or weighing paper should ordinarily *not* be used for weighing reagents since the substance being weighed is easily spilled from such paper.

4. Procedures in this manual are generally written in such terms as "weigh 0.5 grams of substance (to the nearest milligram)." This does not mean that exactly 0.500 grams of substance is needed. The statement means you should obtain an amount of substance between 0.450 and 0.550 grams, and then record the actual amount of substance taken (e.g., 0.496 grams). Unless a procedure states explicitly to weigh out an exact amount (e.g., "weigh out exactly 5.00 grams of NaCl"), you should not waste time trying to obtain an exact amount. However, always record the amount actually taken to the degree of precision possible for the balance.

5. For accurate mass determinations, the object to be weighed must be at room temperature. If a hot or warm object is placed on the pan of the balance, the object causes the air around it to become

heated. Warm air rises; the motion of air may be detected by the balance, giving mass determinations that are significantly less than the true value.

6.   For many types of balances, there are likely to be small errors in the absolute masses of objects determined with the balance, particularly if the balance has not been properly calibrated or has been abused. For this reason, most mass determinations in the laboratory should be performed by a *difference* method: an empty container is weighed on the balance, and then the reagent or object whose mass is to be determined is added to the container. The resulting *difference in mass* is the mass of the reagent or object. Because of possible calibration errors, the same balance should be used throughout a procedure.

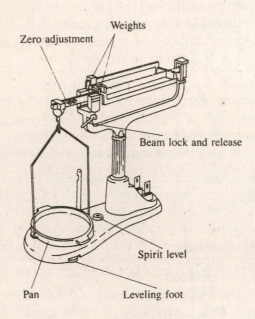
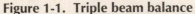

Figure 1-1. Triple beam balance
Never weigh a chemical directly on the pan of the balance, since it is more difficult to "zero" this type of balance if the balance becomes dirty.

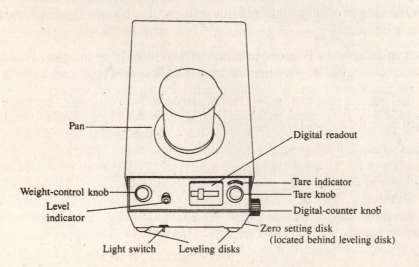

Figure 1-2. Top-loading electric balance
This sort of balance is operated manually: the digits are dialed in by the operator until balance with the object being weighed is achieved.

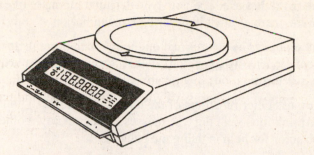

Figure 1-3. Digital electronic balance
The balance directly gives the mass when an object is placed on the pan. The balance must be calibrated and set to zero before use.

Safety Precautions	• Safety eyewear approved by your institution must be worn at all times while you are in the laboratory, whether or not you are working on an experiment

Apparatus/Reagents Required

- unknown mass samples provided by the instructor
- small beakers

Procedure

Record all data and observations directly on the report form in ink.

Examine the balances that are provided in your laboratory. If you are not familiar with the operation of the type of balance available, ask your instructor for a demonstration of the appropriate technique.

Your instructor will provide you with two small objects. You will determine their masses. The objects are coded with an identifying number or letter. Record these identification codes on the report page.

Determine and record the masses of two small beakers that can accommodate the objects whose masses are to be determined. Label the beakers as A and B so you can distinguish them from one another. The determination of the beakers' masses should be to the level of precision permitted by the particular balance you are using.

Transfer the first unknown object to beaker A, and determine the combined mass of beaker A and object. Record. Determine the mass of the unknown object by subtraction. Record.

Transfer the first unknown object to beaker B, and determine the combined mass of beaker B and object. Record. Determine the mass of the unknown object by subtraction. Record.

Although the two beakers used for the determination undoubtedly had different masses when empty, you should have discovered that the mass of the unknown object was the same, regardless of which beaker it was weighed in. Explain.

Repeat the process with the second unknown object and the two beakers.

Using a different balance from that used previously, determine the mass of each of the unknown objects on the second balance in the manner already described, using beaker A only.

Compare the masses of the objects as determined on the two balances. Is there a difference in the masses determined for each object, depending on which balance was used? In future experiments, always use the *same balance* for all mass determinations in a given experiment.

Show the results of your mass determinations of the unknown objects to your instructor, who will compare your mass determinations with the true masses of the unknown objects. If there is any major discrepancy, ask the instructor for help in using the balances.

EXPERIMENT 2

The Use of Volumetric Glassware

Objective

In this experiment, the precision and accuracy permitted by common laboratory volumetric glassware will be examined. The cleaning and care of such glassware will also be discussed.

Introduction

Most of the glassware in your laboratory locker has been marked by the manufacturer to indicate the volume contained by the glassware when filled to a certain level. The graduations etched or painted onto the glassware by the manufacturer differ greatly in the *precision* they indicate, depending on the type of glassware and its intended use. For example, beakers and Erlenmeyer flasks are marked with very *approximate* volumes, which serve merely as a rough guide to the volume of liquid in the container. Other pieces of glassware – notably burets, pipets and graduated cylinders – are marked much more carefully by the manufacturer to indicate precise volumes. It is important to distinguish when a *precise* volume determination is necessary and appropriate for an experiment and when only a *rough* determination of volume is needed.

Glassware that is intended to contain or to deliver specific precise volumes is generally marked by the manufacturer with the letters "TC" (to contain) or "TD" (to deliver). For example, a flask that has been calibrated by the manufacturer to contain exactly 500 mL of liquid at 20°C would have the legend "TC 20 500 mL" stamped on the flask. A pipet that is intended to deliver a precise 10.00 mL sample of liquid at 20°C would be stamped with "TD 20 10 mL." It is important not to confuse "TC" and "TD" glassware: such glassware may not be used interchangeably (for example, a "TD 20 10 mL" pipet may *deliver* 10.00 mL, but may *contain* somewhat more than 10.00 ml when filled, to allow for liquid retained in the tip after dispensing the sample). The temperature (usually 20°C) is specified with volumetric glassware since the volume of a liquid changes with temperature, which causes the density of the liquid to change. While a given pipet will contain or deliver the same *volume* at any temperature, the *mass* (amount of substance present in that volume) will vary with temperature.

A. Graduated Cylinders

The most common apparatus for routine determination of liquid volumes is the *graduated cylinder*. Although a graduated cylinder does not permit as precise a determination of volume as do other volumetric devices, the precision given by the graduated cylinder is sufficient for many applications. Figures 2-1 and 2-2 show typical graduated cylinders. In Figure 2-2, notice the plastic safety ring, which helps to keep the graduated cylinder from breaking if it is tipped over. In Figure 2-1, compare the difference in graduations shown for the 10-mL and 100-mL cylinders. Examine the graduated cylinders in your lab locker, and determine the smallest graduation of volume that can be determined with each cylinder.

9

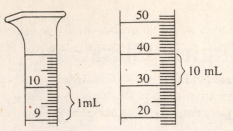

Figure 2-1. Expanded view of 10-mL and 100-mL graduated cylinders
Greater precision is possible with the 10-mL cylinder, since each numbered scale division represents 1 mL.

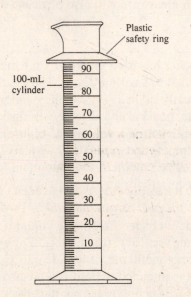

Figure 2-2. A 100-mL graduated cylinder
Note the plastic ring that prevents damage to the cylinder if the cylinder should be tipped over. The safety ring should be near the top during use.

When water (or any aqueous solution) is contained in a narrow glass container such as a graduated cylinder, the liquid surface is not flat. Rather, the liquid surface is curved (see Figure 2-3). This curved surface is called a **meniscus**; it is caused by the interaction between the water molecules and the molecules of the glass container wall. When reading the volume of a liquid that makes a meniscus, hold the graduated cylinder so that the meniscus is at eye level, and read the liquid level at the *bottom* of the curved surface.

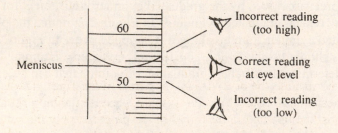

Figure 2-3. Reading a meniscus
Read the bottom of the meniscus while holding at eye level.

B. Pipets

When a more precise determination of liquid volume is needed than can be provided by a graduated cylinder, a transfer *pipet* may be used. Pipets are especially useful if several measurements of the same volume are needed (such as in preparing similar-sized samples of a liquid substance). Two types of pipets are commonly available, as indicated in Figure 2-4. The *Mohr pipet* is calibrated at least to each milliliter and can be used to deliver *any size* sample (up to the capacity of the pipet). The volumetric *transfer pipet* can deliver only *one size* sample (as stamped on the barrel of the pipet), but generally it is easier to use and is more reproducible.

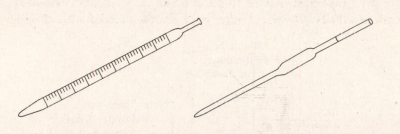

Figure 2-4. A Mohr pipet (left) and a volumetric transfer pipet (right)
Note the calibration mark on the volumetric transfer pipet: the pipet will contain the specified volume when filled exactly to this mark.

Pipets are filled using a **rubber safety bulb** to supply the suction needed to draw liquid into the pipet. *It is absolutely forbidden to pipet by mouth in the chemistry laboratory.* Two common types of rubber safety bulbs are shown in Figure 2-5.

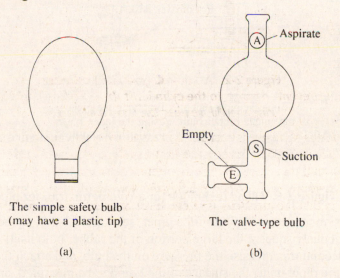

The simple safety bulb
(may have a plastic tip)

The valve-type bulb

(a)   (b)

Figure 2-5. Two common types of pipet safety bulbs: (a) the simple safety bulb (may have a plastic adapter tip); (b) the valve type bulb. *Never pipet by mouth.*

The simple bulb should not actually be placed onto the barrel of the pipet. Doing so would most likely cause the liquid being measured to enter the bulb. Instead, squeeze the bulb first, then merely press the opening of the bulb against the opening in the barrel of the pipet to apply the suction force, keeping the tip of the pipet under the surface of the liquid being sampled. Allow the suction to draw liquid into the pipet until the liquid level is 1 to 2 inches above the calibration mark on the barrel of the pipet. At this point, quickly place your index finger over the opening at the top of the pipet to prevent the liquid level from falling. By gently releasing the pressure of your index finger, the liquid level can be allowed to fall until it reaches the desired calibration mark of the pipet. The tip of the pipet may then be inserted into the container that is to receive the sample and the pressure of the finger removed to allow the liquid to flow from the barrel of the pipet. (See Figure 2-6.) If it is necessary to squeeze the bulb a second time to fill the pipet, remove the bulb from contact with the pipet before squeezing it a second time.

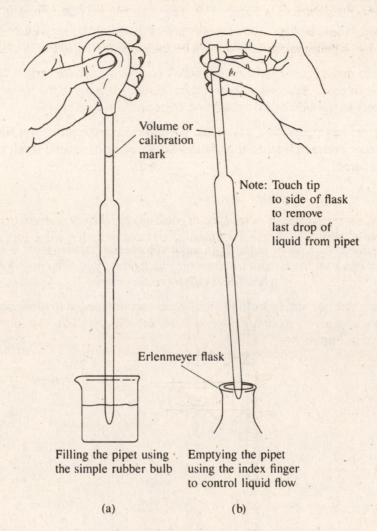

Volume or calibration mark

Note: Touch tip to side of flask to remove last drop of liquid from pipet

Erlenmeyer flask

Filling the pipet using the simple rubber bulb

Emptying the pipet using the index finger to control liquid flow

(a)   (b)

Figure 2-6. Filling technique for a volumetric transfer pipet

To use the more expensive valve-type bulb to fill a pipet (see Figure 2-5b), squeeze the valve of the bulb marked A, and simultaneously squeeze the large portion of the rubber bulb itself to expel air from the bulb. Press valve A a second time, release the pressure on the bulb, and attach the bulb to the top of the pipet. Insert the tip of the pipet under the surface of the liquid to be measured, and squeeze the valve marked S on the bulb, which will cause liquid to begin to be sucked into the pipet. When the liquid level has risen to an inch or two above the calibration mark of the pipet, stop squeezing valve S to stop the suction. Transfer the pipet to the vessel to receive the liquid, and press valve E to empty the pipet. The use of this sort of bulb generally requires considerable practice to develop proficiency.

When using either type of pipet, observe the following rules:

1.   The pipet must be scrupulously clean before use: wash with soap and water, rinse with tap water and then with distilled water. If the pipet is clean enough for use, water will not bead up anywhere on the inside of the barrel.

2. To remove rinse-water from the pipet (which will prevent dilution of the solution to be measured), rinse the pipet with several small portions of the solution to be measured, discarding the rinsings in a waste beaker for disposal. It is not necessary to completely fill the pipet with the solution for rinsing. Draw 2 to 3 mL into the pipet, then place your fingers over both ends of the pipet, hold the pipet horizontally, and rotate the pipet several times to wash the walls of the pipet with the liquid.

3. The tip of the pipet must be kept *under the surface* of the liquid being measured out during the entire time suction is being applied, or air will be sucked into the pipet.

4. Allow the pipet to drain for at least a minute when emptying to make certain the full capacity of the pipet has been delivered. Remove any droplets of liquid adhering to the tip of the pipet by touching the tip of the pipet to the side of the vessel that is receiving the sample.

5. If you are using the same pipet to measure out several different liquids, you should rinse the pipet with distilled water between liquids, then follow with a rinse of several small portions of the next liquid to be measured.

C. Burets

When samples of various sizes must be dispensed or measured precisely, a buret may be used. The buret consists of a tall, narrow calibrated glass tube, fitted at the bottom with a valve for controlling the flow of liquid. The valve is more commonly called a **stopcock**. (See Figure 2-7.)

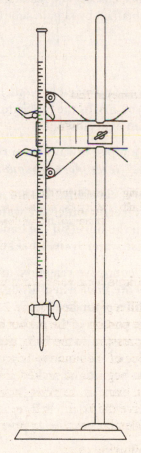

Figure 2-7. A typical student buret
Most commonly, 50-mL burets are used in chemistry labs.

Like a pipet, a buret must be scrupulously clean before use. The precision permitted in reading a buret is on the order of 0.02 mL, but if the buret is not completely clean, this level of precision is not attainable. The buret should be cleaned at your bench, not in the sink (most sinks are too small for the length of the buret, and the buret is very easily broken). To clean the buret, first fill the buret with a small amount of soap and water using a funnel, then use a special long-handled buret brush to scrub the interior of the glass. Finally, rinse the buret by pouring tap water through it (allowing the water to drain from the tip), followed by several rinsings with distilled water. Do not try to clean or rinse the pipet directly from the faucet.

Before use, the buret should be rinsed with several small portions of the solution to be used in the buret. The buret should be tilted and rotated during the rinsings, to make sure that all rinse water is washed from it. Discard the rinsings. After use, the buret should again be rinsed with distilled water. Many of the reagent solutions used in burets may attack the glass of the buret if they are not removed, destroying the calibration of the buret. To speed up the cleaning of a buret in future experiments, the buret may be left filled with distilled water during storage between experiments (if your locker is large enough to permit it).

A common mistake made by beginning students is to fill the buret with the reagent solution to be dispensed to exactly the 0.00 mL mark. This is *not* necessary or desirable in most experiments, and it wastes time. The buret should be filled to a level that is comfortable for you to read (based on your height). The precise initial liquid level reading of the buret should be taken *before* the solution is dispensed and again *after* the liquid has been dispensed. The readings should be made to the nearest 0.02 mL. The volume of liquid dispensed is then obtained by simple subtraction of the two volume readings.

Safety Precautions	• **Safety eyewear approved by your institution must be worn at all times while you are in the laboratory, whether or not you are working on an experiment** • **When using a pipet, always use a rubber safety bulb to apply the suction force.** ***Never pipet by mouth.*** • **Rinse the buret carefully. Do not attempt to place the top opening of the buret directly under the water tap: this usually breaks the buret. Set up the buret on the bench using a buret clamp/stand, and pour rinse water from a beaker into the buret using a funnel.** • **With caustic/corrosive reagents, use a funnel and place the buret and stand *on the floor* when filling. Never fill a buret whose opening is above your head.**

Apparatus/Reagents Required

- graduated cylinders
- pipets and safety bulb
- buret and clamp
- beakers
- distilled water

Procedure

Record all data and observations directly on the report page in ink.

1. The Graduated Cylinder

Your instructor will set up a display of several graduated cylinders filled with different amounts of colored water. There are several sizes of cylinder available (10-mL, 25-mL, 50-mL and 100-mL). Examine each cylinder, paying particular attention to the marked scale divisions on the cylinder. For each graduated cylinder, to what fractional unit of volume does the smallest scale mark correspond?

Read the volume of liquid contained in each graduated cylinder and record it. Make your readings to the level of precision permitted by each of the cylinders. Remember to read the liquid level as tangent to the meniscus formed by the liquid.

Check your readings of the liquid levels with the instructor before proceeding, and ask for assistance if your readings differ from those provided by the instructor.

Clean and wipe dry your 25-mL graduated cylinder and a 50-mL beaker (a rolled-up paper towel will enable you to dry the interior of the graduated cylinder). Weigh (separately) the graduated cylinder and the beaker, and record the mass of each to the nearest milligram (0.001 gram).

Obtain about 100 mL of distilled water in a clean Erlenmeyer flask. Determine and record the temperature of the distilled water with a thermometer.

Fill the graduated cylinder with distilled water so that the meniscus of the water level lines up with the 25-mL calibration mark of the cylinder. Place distilled water in the 50-mL beaker up to the 25-mL mark. A clean plastic pipet may be used to add small portions of water to the cylinder to get the water level to exactly the calibration mark.

Weigh (separately) the graduated cylinder and the 50-mL beaker to the nearest milligram (0.001 gram) and calculate the *mass* of water that each contains.

Using the Table of Densities for water from the Appendix to this manual, calculate the *volume* of water present in the graduated cylinder and beaker from the exact *mass* of water present in each.

Compare the *calculated* volume of water (based on the mass of water) to the *observed* volumes of water determined from the calibration marks on the cylinder and beaker. Calculate the percentage *difference* between the calculated volume and the observed volume from the calibration marks. Why are the calibration marks on graduated cylinders and beakers taken only to be an approximate guide to volume?

2. The Pipet

Obtain a 25-mL pipet and rubber safety bulb. Clean the pipet with soap solution: draw a few mL of soap solution into the pipet, place your index fingers over the openings in either end of the pipet, hold the pipet horizontally, and rotate the pipet in your hand so as to coat all the inner surfaces with the soap solution. Rinse the pipet with tap water and then with small portions of distilled water. Practice filling and dispensing distilled water from the pipet using the rubber safety bulb until you feel comfortable with the technique. Ask your instructor for assistance if you have any difficulties.

Clean and wipe dry a 150-mL beaker. Weigh the beaker to the nearest milligram (0.001 gram) and record.

Obtain about 100 mL of distilled water in a clean Erlenmeyer flask. Determine and record the temperature of the water.

Pipet exactly 25 mL of the distilled water from the Erlenmeyer flask into the clean beaker you have weighed. Reweigh the beaker containing the 25 mL of water. Determine the mass of water transferred by the pipet.

Using the Table of Densities found in the Appendix to this manual, calculate the *volume* of water transferred by the pipet from the *mass* of water transferred. Compare this calculated volume to the volume of the pipet as specified by the manufacturer. Any significant difference in these two volumes is an indication that you need additional practice in pipeting. Consult with your instructor for help.

How does the volume dispensed by the pipet compare to the volumes as determined in Part 1 using a graduated cylinder or beaker?

3. The Buret

Obtain a buret and set it up in a clamp attached to a ring stand on your lab bench. Then move the buret and ring stand to the floor. Place a large beaker under the stopcock/delivery tip of the buret.

Using a funnel, fill the buret with tap water, and check to make sure that there are no leaks from the stopcock before proceeding. If the stopcock leaks, have the instructor examine the stopcock to make sure that all the appropriate washers are present. If the stopcock cannot be made leak-proof, replace the buret.

Add a few mL of soap solution to the buret and use a long-handled buret brush to scrub the inner surface of the buret. Rinse all soap from the buret with tap water, being sure to flush water through the stopcock as well. Rinse the buret with several small portions of distilled water.

Place the buret and its stand on the floor. Then, using a funnel, fill the buret to above the zero mark with distilled water. Return the buret and its stand to your workbench.

Open the stopcock of the buret and allow the distilled water to run from the buret into a beaker or flask. Examine the buret while the water is running from it. If the buret is clean enough for use, water will flow in sheets down the inside surface of the buret without beading up anywhere. If the buret is not clean, repeat the scrubbing with soap and water.

Once the buret is clean, refill it with distilled water to a point somewhat below the zero mark. Determine the precise liquid level in the buret to the nearest 0.02 mL.

With a paper towel, clean and wipe dry a 150-mL beaker. Determine the mass of the beaker to the nearest milligram (0.001 g).

Place the weighed beaker beneath the stopcock of the buret. Open the stopcock of the buret and run water into the beaker until approximately 25 mL of water have been dispensed. Determine the precise liquid level in the buret to the nearest 0.02 mL. Calculate the volume of water that has been dispensed from the buret by subtraction of the two liquid levels.

Reweigh the beaker (containing the water dispensed from the buret) to the nearest milligram, and determine the mass of water transferred to the beaker from the buret.

Use the Table of Densities in the Appendix to calculate the volume of water transferred, using the mass of the water. Compare the volume of water transferred (as determined by reading the buret) with the calculated volume of water (from the mass determinations). If there is any significant difference between the two volumes, you most likely need additional practice in the operation and reading of the buret.

How does the volume dispensed by the buret compare to the volumes as determined in Part 1 using a graduated cylinder or beaker? How does the volume dispensed by the buret compare to that dispensed using a pipet in Part 2?

EXPERIMENT 3

The Measurement of Temperature

Objective

In this experiment, you will check your thermometer for errors by determining the temperatures of two stable reference equilibrium systems. You will then use your calibrated thermometer in determining the boiling point of an unknown substance.

Introduction

The most common laboratory device for the measurement of temperature is, of course, the thermometer. The typical thermometer used in the general chemistry laboratory permits the determination of temperatures from –20° to 120°C. Most laboratory thermometers are constructed of glass, and so they are very fragile. Mercury is used as the temperature sensing fluid in many thermometers, and if the thermometer is broken, poisonous mercury may be released. *Any mercury spills must be reported immediately to the instructor.* Because of the dangers of mercury, other liquids (such as colored alcohol) are beginning to be used more commonly in student-grade laboratory thermometers.

The typical laboratory thermometer contains a bulb (reservoir) of mercury or other liquid at the bottom; it is this portion of the thermometer that actually senses the temperature. The glass barrel of the thermometer above the liquid bulb contains a fine capillary opening in its center, into which the liquid rises as it expands in volume when heated. The capillary tube in the barrel of the thermometer has been manufactured to very strict tolerances, and it is very regular in cross-section along its length. This regularity ensures a direct relationship between the rise in the level of liquid in the capillary tube and the temperature of the thermometer's surroundings.

Although the laboratory thermometer may appear similar to the sort of clinical thermometer used for determination of body temperature, the laboratory thermometer does *not* have to be shaken before use. Medical thermometers are manufactured with a constriction in the capillary tube that is intended to prevent the liquid level from changing once it has risen. The liquid level of a laboratory thermometer, however, changes immediately when removed from the substance whose temperature is being measured. For this reason, temperature readings with the laboratory thermometer must be made while the bulb of the thermometer is immersed in the material being determined.

Because the laboratory thermometer is so fragile, it is helpful to check that the thermometer provides reliable readings before any important determinations are made with it. Often, thermometers develop nearly invisible hairline cracks along the barrel, making them unsuitable for further use. These breaks happen if you are not careful in opening and closing your laboratory locker.

To check whether or not your thermometer is reading temperatures correctly, you will *calibrate* the thermometer. To do this, you will determine the reading given by your thermometer in two systems whose temperature is known with certainty. If the readings given by your thermometer differ by more than one degree from the true temperatures of the systems measured, you should exchange your thermometer and calibrate the new thermometer. A mixture of ice and water has an equilibrium temperature of exactly 0°C, and will be used as the first calibration system. A boiling water bath (whose exact temperature can be determined from the day's barometric pressure) will be used as the second calibration system in this experiment.

Once your thermometer has been calibrated, you will use the thermometer in a simple but very important experiment: the determination of the boiling point of a pure chemical substance. The boiling points of pure substances are important in chemistry because they are "characteristic" for a given substance; under the same laboratory conditions, a given substance will always have the same boiling point. Characteristic physical properties, such as the boiling point of a substance, are of immense help in the identification of unknown substances. Such properties are routinely reported in scientific papers when new substances are isolated or synthesized, and are compiled in tables in the various handbooks of chemical data that are available in science libraries. When an unknown substance is isolated from a chemical system, its boiling point may be measured (along with other characteristic properties) and then compared with tabulated data. If the experimentally determined physical properties of the unknown substance match those found in the literature, you may typically assume that you have identified the unknown substance.

The boiling point of a liquid is defined as the temperature at which the vapor escaping from the surface of the liquid has a pressure equal to the pressure existing above the liquid. In the most common situation of a liquid boiling in a container open to the atmosphere, the pressure above the liquid will be the day's barometric pressure. In other situations, the pressure above a liquid may be reduced by means of a vacuum pump or aspirator, which enables the liquid to be boiled at a much lower temperature than in an open container (this is especially useful in chemistry when a liquid is unstable, possibly decomposing if it were heated to its normal boiling point under atmospheric pressure).

When boiling points are tabulated in the chemical literature, the pressure at which the boiling point determinations were made is also listed. The method to be used for the determination of boiling point is a semi-micro method that requires only a few drops of liquid.

Apparatus/Reagents Required

- thermometer and clamp
- several beakers
- melting point capillaries
- 10 × 75 mm test tubes
- burner and rubber tubing
- glass file
- medicine dropper
- unknown sample for boiling point determination
- ice

Safety Precautions	• Safety eyewear approved by your institution must be worn at all times while you are in the laboratory, whether or not you are working on an experiment. • Thermometers are often fitted with rubber stoppers as an aid in supporting the thermometer with a clamp. Inserting a thermometer through a rubber stopper must be done carefully to prevent breaking of the thermometer, which might cut you. Your instructor will demonstrate the proper technique for inserting your thermometer through the hole of a rubber stopper. Glycerine is used to lubricate the thermometer and stopper. Protect your hands with a towel during this procedure. • Mercury is poisonous and is absorbed through the skin. Its vapor is toxic. If mercury is spilled from a broken thermometer, inform the instructor immediately so that the mercury can be removed. Do not attempt to clean up mercury yourself. • The liquids used in the experiment are flammable. Although only small samples of the liquids are used, the danger of fire is not completely eliminated. Keep the liquids away from all open flames! • The liquids used in this experiment are toxic if inhaled or absorbed through the skin. The liquids should be disposed of in the manner indicated by the instructor.

Procedure

Record all data and observations directly on the report page in ink.

1. Calibration of the Thermometer

Fill a 400-mL beaker with ice, and add tap water to the beaker until the ice is covered with water. Stir the mixture with a stirring rod for 30 seconds.

Clamp the thermometer to a ring stand so that the bottom 2–3 inches of the thermometer is dipping into the ice bath. Make certain that the thermometer is suspended freely in the ice bath and is *not* touching either the walls or the bottom of the beaker.

Allow the thermometer to stand in the ice bath for 2 minutes, and then read the temperature indicated by the thermometer to the nearest 0.2°C. Remember that the thermometer must be read while still in the ice bath. If the reading indicated by the thermometer differs from 0°C by more than one degree, replace the thermometer and repeat the ice bath calibration.

Allow the thermometer to warm to room temperature by resting it in a safe place on the laboratory bench. Set up the apparatus for boiling as indicated in Figure 3-1, using a 100-mL beaker containing approximately 75 mL of water. Add two or three boiling chips to the water, and heat the water to boiling.

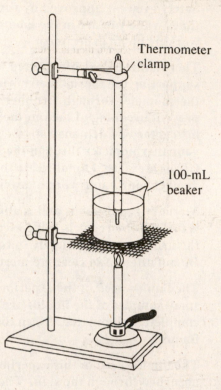

Thermometer clamp

100-mL beaker

Figure 3-1. Apparatus for calibration of the thermometer
Make certain the thermometer is suspended freely in the water bath and does not touch the bottom of the beaker.

Using a clamp, suspend the thermometer so that it is dipping halfway into the boiling water bath. Make certain that the thermometer is *not* touching the walls or bottom of the beaker. Allow the thermometer to stand in the boiling water for 2 minutes; then record the thermometer reading to the nearest 0.2°C.

A boiling water bath has a temperature near 100°C, but the actual temperature of boiling water is dependent on the barometric pressure and changes with the weather from day to day. Your instructor will list the current barometric pressure on the chalkboard. Using a handbook of chemical data, look up the actual boiling point of water for this barometric pressure and record it.

If your measured boiling point differs from the handbook value for the provided barometric pressure by more than one degree, exchange your thermometer at the stockroom, and repeat the calibration of the thermometer in both the ice bath and the boiling water bath.

2. Determination of Boiling Point of an Unknown Liquid

Set up a beaker half-filled with water on a wire gauze on an iron ring/ring stand.

Obtain a clean, dry, 10×75-mm semi-micro test tube, which will contain the boiling-point sample. Attach the test tube to the lower end of your thermometer with two small rubber bands or thinly cut rings of rubber tubing. See Figure 3-2 on the following page.

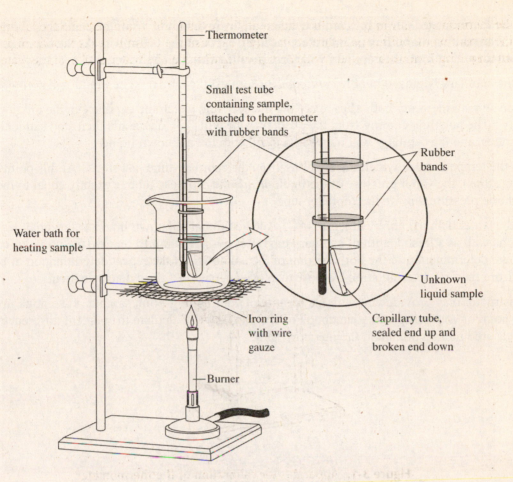

Figure 3-2. Apparatus for boiling-point determination

Obtain an unknown liquid for boiling-point determination from your instructor, and record the identification code number of the unknown in your notebook and on the lab report page. Transfer part of the unknown sample to the small test tube, until the test tube is approximately half full.

Obtain a melting-point capillary. If the capillary tube is open at both ends, heat one end briefly in a flame to seal it off. Using a glass file, carefully cut the capillary about 1 inch from the sealed end. *Caution!* Do *not* fire-polish the cut end of the capillary.

Place the small portion of capillary, *sealed end facing up,* into the boiling-point sample in the test tube.

The capillary has a rough edge at the cut end, which serves as a surface at which bubbles can form during boiling. The capillary is filled with air when inserted, sealed end up, into the liquid, and the presence of this air can be used to judge when the vapor pressure of the unknown liquid reaches atmospheric pressure.

Lower the thermometer and sample into the water bath so that the bulb of the thermometer and the sample are suspended freely and not touching the bottom or sides of the beaker. Place a stirring rod in the beaker.

Begin heating the water bath in the beaker with a low flame so that the temperature rises by 1 or 2°C per minute. Stir the water bath with the stirring rod to make sure the heat being applied is distributed evenly. Watch the small capillary tube in the unknown sample while heating.

As the sample is heated, air in the capillary tube will begin to bubble from the capillary. As the air bubbles from the capillary, it is gradually replaced by vapor of the unknown. As the liquid begins to boil, the bubbles coming from the capillary will form a continuous, rapid stream.

When the capillary begins to bubble *continuously*, turn off the burner. The liquid will continue to boil.

Continue stirring the water bath after removing the heat, and watch the capillary in the unknown sample carefully. The bubbling coming from the capillary tube will slow down and then stop suddenly after a few minutes, and the capillary will then begin to fill with the unknown liquid.

Record the temperature at which the bubbling from the capillary tube just stops. At this point (where the bubbling stops), the vapor pressure of the liquid *inside* the capillary tube is exactly equal to the atmospheric pressure *outside* the capillary tube.

Allow the water bath to cool by at least 20°C. Add additional unknown liquid to the small test tube if needed, as well as a fresh length of capillary tube (it is not necessary to remove the previous capillary). Repeat the determination of the boiling point of the unknown. If the repeat determination of boiling point differs from the first determination by more than one degree, do a third determination.

Your instructor may provide you with the identity of your boiling-point sample. If so, look up the true boiling point of your sample in a handbook of chemical data. Calculate the percent difference between your measured boiling point and the literature value.

EXPERIMENT 4

Recrystallization and Melting Point Determination

Objective

The separation of mixtures into their constituent components defines an entire subfield of chemistry referred to as **separation science**. In this experiment, **recrystallization**, one of the most common techniques for the resolution of mixtures of solids, will be examined.

Introduction

Mixtures occur very commonly in chemistry. When a new chemical substance is synthesized in a research lab, for example, the new substance usually must be separated from various side-products, catalysts, and any excess starting reagents still present. When a substance needs to be isolated from a natural biological source, the substance of interest is generally found in a very complex mixture with many other substances, all of which must be removed. Chemists have developed a series of standard methods for resolution and separation of mixtures, one of which – recrystallization – will be investigated in this experiment.

Mixtures of solids may often be separated on the basis of differing *solubilities* of the components. If one of the components of the mixture is very soluble in water, for example, while the other components are insoluble, the water-soluble component may be removed from the mixture by simple filtration through ordinary filter paper. A more general case occurs when all the components of a mixture are soluble, but to different extents, in water or some other solvent. The solubility of substances in many cases is greatly influenced by temperature. By controlling carefully the temperature at which solution occurs or at which a filtration is performed, it may be possible to separate the components of the mixture. Most commonly, a sample is added to the solvent and is then heated to boiling. The hot solution is then filtered to remove completely insoluble substances. The sample is then cooled, either to room temperature or below, which causes recrystallization of those substances whose solubilities are very temperature-dependent. These crystals can then be isolated by filtration, and the filtrate remaining can be concentrated to reclaim those substances whose solubilities are not so temperature-dependent.

After a substance has been isolated as a pure solid from a mixture, it is a very common practice to determine the **melting point** of the material. When a pure solid melts during heating, the melting usually occurs quickly at one specific, *characteristic* temperature. For certain substances, especially more complicated organic substances or biological substances that may decompose slightly when heated, the melting may occur over a span of several degrees, called the **melting range**. Melting ranges are also commonly observed if the substance being determined is not completely pure. The presence of an impurity will broaden the melting range of the major component and will also lower the temperature at which melting begins. As with boiling points (Experiment 3), the characteristic melting points of pure, solid substances are routinely reported in the scientific literature and are tabulated in handbooks for use in the subsequent identification of unknown substances.

Safety Precautions	• Safety eyewear approved by your institution must be worn at all times while you are in the laboratory, whether or not you are working on an experiment. • The solid mixture contains benzoic acid, which may be irritating to the skin and respiratory tract. It is poisonous if ingested. • When moving hot containers, use metal tongs or a towel to avoid burns. Beware of burns from steam while solutions are being heated. • *Caution!* Oil is used as the heating fluid in the Thiele tube used for the melting-point determinations. Hot oil may spatter if it is heated too strongly, especially if any moisture is introduced into the oil from glassware that is not completely dry. The oil may smoke or ignite if heated above 200°C. <u>If the oil appears cloudy, inform the instructor and do not heat it.</u> • Dispose of all solids and liquids as directed by the instructor.

Apparatus/Reagents Required

- impure benzoic acid sample (benzoic acid that has been colored with charcoal)
- melting point capillaries
- filter paper
- rubber bands
- oil-filled Thiele tubes

Procedure

Record all observations and data directly on the report page in ink.

1. Recrystallization

Obtain a sample of impure benzoic acid for recrystallization. Benzoic acid is fairly soluble in hot water, but has a much lower solubility in cold water. The benzoic acid has been contaminated with charcoal, which is not soluble under either temperature condition. Transfer the benzoic acid sample to a clean 150-mL beaker.

Set up a short-stem gravity filter funnel in a small metal ring clamped to a ring stand. Fit the filter funnel with a piece of filter paper folded in quarters to make a cone. See Figure 4-1(a).

Moisten the filter paper slightly so that it will remain in the funnel. Place a clean 250-mL beaker beneath the stem of the funnel.

Set up a second 250-mL beaker, about half filled with distilled water, on a wire gauze over a metal ring. Heat the water to boiling.

When the water is boiling, pour about two-thirds of the water into the beaker containing the benzoic acid sample. Use a towel to protect your hands from the heat and steam.

Pour the remainder of the boiling water through the gravity funnel to heat it. If the funnel is not preheated, the benzoic acid may crystallize in the stem of the funnel rather than passing through it. Discard the water that was used to preheat the funnel.

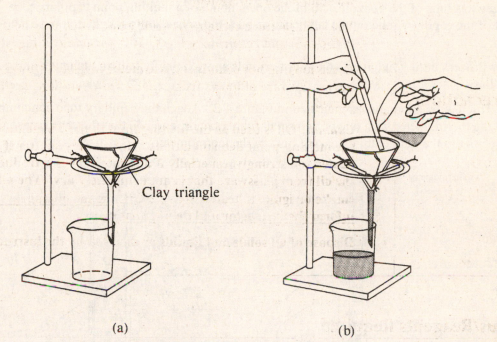

(a)   (b)

Figure 4-1. Filtration of a hot solution
Use a stirring rod as a guide for running the hot solution into the funnel. Do not fill the funnel more than half-full at a time, to prevent the solution being lost over the rim of the filter paper cone.

Transfer the 150-mL beaker containing the benzoic acid mixture to the burner and reheat it until the mixture just begins to boil again (this should only take a few seconds). Stir the mixture to make sure that the benzoic acid dissolves to the greatest extent possible.

Using a towel to protect your hands, pour the benzoic acid mixture through the preheated funnel. Catch the filtrate in a clean beaker. See Figure 4-1(b).

Allow the filtrate to cool to room temperature.

When the benzoic acid solution has cooled to room temperature, filter the crystals to remove water. Wash the crystals with two 10-mL portions of cold water.

Transfer the filtrate from which the crystals have been removed to an ice bath to see if additional crystals will form at the lower temperature. Examine, but do not isolate this second crop of crystals.

Transfer the filter paper containing the benzoic acid crystals to a watch glass, and heat the watch glass over a 400-mL beaker of boiling water to dry the crystals. You can monitor the drying of the crystals by watching for the filter paper to dry out as it is heated on the water bath.

When the benzoic acid is completely dry, determine the melting point of the recrystallized material using the method discussed in Part 2.

2. Melting Point Determination

If the benzoic acid recrystallized in Part 1 is not finely powdered, grind some of the crystals on a watch glass or flat glass plate with the bottom of a clean beaker. Your instructor may also have mortars and pestles available for grinding.

Pick up a few crystals of the benzoic acid in the open mouth of a melting point capillary tube. Tap the sealed end of the capillary tube on the lab bench to pack the crystals into a tight column at the sealed end of the tube.

Repeat this process until you have a column of crystals approximately half an inch high at the sealed end of the capillary.

Set up the Thiele tube apparatus as indicated in Figure 4-2. Attach the capillary tube containing the crystals to the thermometer with one or two small rubber bands, and position the capillary so that the crystals are next to the temperature-sensing bulb of the thermometer.

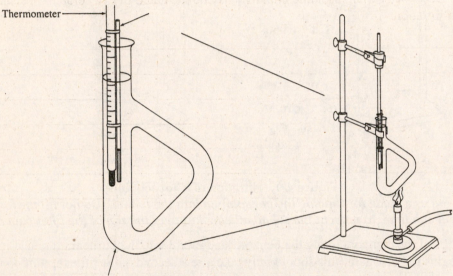

Thermometer

Thiele tube (oil-filled)

Figure 4-2.   Thiele tube oil bath for melting point determinations
Exercise caution when dealing with hot oil. Replace the oil if it appears at all cloudy or burned.

Lower the thermometer into the oil bath, and begin heating the bottom of the side arm of the Thiele tube with a very small flame. Adjust the flame as necessary so that the temperature rises only by 1 or 2°C per minute.

Watch the crystals in the capillary tube, and record the *exact temperature* at which the benzoic acid crystals first begin to melt, and the exact temperature at which the last portion finishes melting. Record these two temperatures as the melting range of the benzoic acid.

Allow the oil bath to cool by at least 20°C.

Prepare another sample of the recrystallized benzoic acid in a fresh capillary tube, and repeat the determination of the melting point. If this second determination differs significantly from the first determination, repeat the experiment a third time.

Look up the melting point of benzoic acid online or in a handbook of chemical data. Compare the melting point of your benzoic acid with that indicated in your reference material. If the melting point you obtain is significantly lower than that reported in your reference, heat the crystals for an additional period over the hot water bath to dry them further, then repeat the melting point determination.

Name: _____ Section: _____

Lab Instructor: _____ Date: _____

EXPERIMENT 4

Recrystallization and Melting Point Determination

Pre-Laboratory Questions

1. In this experiment, you will purify a benzoic acid sample to which some charcoal has been added as an impurity. Describe how an insoluble impurity may be separated from a soluble substance by the process of filtration.

   _____

   _____

   _____

   _____

   _____

   _____

   _____

   _____

   _____

2. Using the Internet or a handbook of chemical data, look up the melting point of pure benzoic acid.

   Melting point _____ °C   Reference _____

3. Recrystallization is a common method applied to separate the components in mixtures of substances, especially to remove a relatively minor impurity from an otherwise pure substance. Use your textbook or an encyclopedia of chemistry to find three other methods of separating mixtures and describe briefly those methods.

   _____

   _____

   _____

   _____

   _____

   _____

35

4. What precautions should be taken when using oil as a heating fluid in a Thiele tube? Why should the oil in the Thiele tube be replaced if it appears cloudy?

_____

_____

_____

_____

_____

_____

5. Why do we use only a very small flame to heat the Thiele tube when determining the melting point of a crystalline solid?

_____

_____

_____

_____

_____

_____

_____

6. When determining the melting point of a sample, if the sample melts over a range of several degrees, what might this imply about the purity of the sample?

_____

_____

_____

_____

_____

_____

_____

Name: _____ Section: _____

Lab Instructor: _____ Date: _____

EXPERIMENT 4

Recrystallization and Melting Point Determination

Results/Observations

1. Recrystallization

Appearance of impure benzoic acid sample

_____

_____

Observations of recrystallization process

_____

_____

Appearance of purified benzoic acid sample

_____

_____

2. Melting Point Determination

Observations of melting point determination

_____

_____

Melting points determined

First sample   _____

Second sample   _____

(Third sample)   _____

Average melting point _____

Error in melting point _____

Questions

1. It was necessary for you to preheat the funnel used for the gravity filtration of the benzoic acid sample. Why?

    _____

    _____

    _____

    _____

2. Why should a melting point capillary be positioned directly next to the temperature-sensing bulb of the thermometer when performing a melting-point determination?

    _____

    _____

    _____

    _____

    _____

    _____

3. Why was it necessary to allow the oil bath to cool by at least 20°C before performing the second determination of the melting point of your recrystallized benzoic acid?

    _____

    _____

    _____

    _____

    _____

4. Why was it necessary to dry the benzoic acid crystals before determining their melting point? What would have happened to the melting point if the crystals had still been wet?

    _____

    _____

    _____

    _____

    _____

    _____

EXPERIMENT 5

Density Determinations

Objective

Density is an important property of matter that may be useful as a method of identification. In this experiment, you will determine the densities of regularly and irregularly shaped solids, as well as the densities of pure liquids and solutions.

Introduction

The density of a sample of matter represents the mass contained within a unit volume of space within the sample. For most samples, a "unit volume" means 1.0 mL. The units of density, therefore, are quoted in terms of grams per milliliter (g/mL) or grams per cubic centimeter (g/cm^3) for most samples of matter.

Since we seldom deal with exactly 1.0 mL of substance in the chemistry laboratory, we usually say that the density of a sample represents the mass of the sample divided by its volume.

$$\text{density} = \frac{\text{mass}}{\text{volume}}$$

Because the density does in fact represent a ratio, the mass of *any* size sample divided by the volume of the sample gives the mass that 1.0 mL of the same sample would possess.

Densities are usually determined and reported at 20°C (around room temperature) because the volume of a sample, and hence the density, will often vary with temperature. This variation is especially present in gases, with smaller (but still often significant) changes for liquids and solids. References (such as the various chemical handbooks) always specify the temperature at which a density was determined.

Density can be used to determine the concentration of solutions in certain instances. When a solute is dissolved in a solvent, the density of the solution will be different from that of the pure solvent itself. Handbooks list detailed information about the densities of solutions as a function of their composition. If a sample is known to contain only a single solute, the density of the solution can be measured experimentally and then the handbook can be consulted to determine what concentration of solute gives rise to the measured density.

The determination of the density of certain physiological liquids is often an important screening tool in medical diagnosis. For example, if the density of urine differs from normal values, the kidneys may be secreting substances that should not be lost from the body. The determination of density (specific gravity) is almost always performed during a urinalysis.

There are several techniques used for the determination of density. The method used will depend on the type of sample and on the precision desired for the measurement. For example, devices that permit a quick, reliable, routine determination have been constructed for determinations of the density of urine. In general, a density determination will involve the determination of the mass of the sample with a balance, but the method used to determine the volume of the sample will differ from situation to situation. Several methods of volume determination are explored in this experiment.

For solid samples, there may be different methods needed for the determination of the volume, depending on whether or not the solid is regularly-shaped. If a solid has a regular shape (e.g., cube, rectangle, cylinder), the volume of the solid may be determined by geometry:

For a cubic solid, volume = $(\text{edge})^3$

For a rectangular solid, volume = length × width × height

For a cylindrical solid, volume = $\pi \times (\text{radius})^2 \times \text{height}$

If a solid does *not* have a regular shape, it may be possible to determine the volume of the solid making use of **Archimedes' principle**, which states that an insoluble, non-reactive solid will *displace* a volume of liquid equal to its own volume. Typically, an irregularly shaped solid is added to a liquid in a volumetric container (such as a graduated cylinder) and the change in liquid level is determined.

For liquids, very precise values of density may be determined by pipeting an exact volume of liquid into a sealable weighing bottle (this technique is especially useful for highly volatile liquids) and then determining the mass of liquid that was pipeted. A more convenient method for routine density determinations for liquids is to weigh a particular volume of liquid as contained in a graduated cylinder.

Safety Precautions	• **Safety eyewear approved by your institution must be worn at all times while you are in the laboratory, whether or not you are working on an experiment.** • **The unknown liquids may be flammable and their vapors may be toxic. Keep the unknown liquids away from open flames and do not inhale their vapors. Dispose of the unknown liquids as directed by the instructor.** • **Dispose of the metal samples only in the special container designated for their collection. Do not combine different metals in the same container.**

Apparatus/Reagents Required

* unknown liquid sample
* unknown metal samples
* sodium chloride
* ruler or calipers

Procedure

Record all data and observations directly on the report page in ink.

1. Determination of the Density of Solids

Obtain a regularly shaped solid, and record its identification number. With a ruler or calipers, determine the physical dimensions of the solid to the nearest 0.2 mm. Calculate the volume of the solid. Be sure to indicate the geometry formula used for the calculation.

Determine the mass of the regularly shaped solid to at least the nearest mg (0.001 g). Calculate the density of the solid.

Obtain a sample of unknown metal pellets (metal shot) or an irregularly shaped chunk of metal, and record its identification code number. Determine the mass of a sample of the metal of approximately 50 g, but record the *actual* mass of metal taken to the nearest mg (0.001 g).

Add water to your 100-mL graduated cylinder – up to approximately the 50-mL mark. Record the exact volume of water in the cylinder to the precision permitted by the calibration marks of the cylinder.

Pour the metal sample into the graduated cylinder, making sure that none of the pellets sticks to the walls of the cylinder above the water level. Stir/shake the cylinder to make certain that no air bubbles have been trapped among the metal pellets. (See Figure 5-1.)

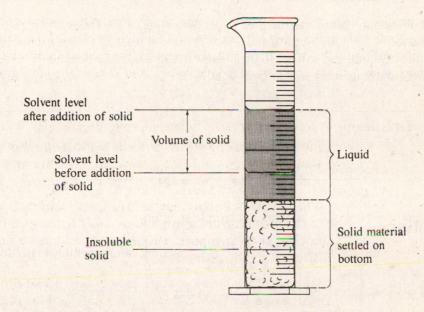

Figure 5-1. Measurement of volume by displacement
A non-soluble object displaces a volume of liquid equal to its own volume.

Read the level of the water in the graduated cylinder, again making your determination to the precision permitted by the calibration marks of the cylinder. Assuming that the metal sample does not dissolve in or react with water, the change in water levels represents the volume of the metal pellets.

Calculate the density of the unknown metal pellets. Turn in the sample of metal pellets to your instructor (*do not discard*).

2. Density of Pure Liquids

Clean and dry a 25-mL graduated cylinder (a rolled-up paper towel may be used to dry the interior of the cylinder). Determine the mass of the dried cylinder to the nearest mg (0.001 g).

Add distilled water to the cylinder so that the water level is *above* the 20-mL mark but *below* the 25-mL mark. Determine the temperature of the water in the cylinder.

Determine the mass of the cylinder and water to the nearest milligram.

Record the exact volume of water in the cylinder, to the correct level of precision permitted by the calibration marks on the barrel of the cylinder.

Calculate the density of the water. Compare the measured density of the water with the value listed online or in a handbook of chemical data for the temperature of your experiment.

Clean and dry the graduated cylinder.

Obtain an unknown liquid and record its identification number. Determine the density of the liquid, using the method just described for water.

3. Density of Solutions

The concentration of a solution is often conveniently described in terms of the solution's percentage composition on a mass basis. For example, a 5% sodium chloride solution contains 5 g of sodium chloride in every 100 g of solution (which corresponds to 5 g of sodium chloride for every 95 g of water present).

Prepare solutions of sodium chloride in distilled water consisting of the following approximate percentages by mass: 5%, 10%, 15%, 20% and 25%. Prepare at least 25 mL of each solution. Make the mass determinations of solute and solvent to the nearest milligram, and calculate the exact percentage composition of each solution based on this mass data.

Example:

Suppose it is desired to prepare 50. g of 7.5% calcium chloride solution. When we say we have a 7.5% calcium chloride solution, we are implying this conversion factor

$$\frac{7.5 \text{ g calcium chloride}}{100. \text{ g solution}} = \frac{7.5 \text{ g calcium chloride}}{7.5 \text{ g calcium chloride} + 92.5 \text{ g water}}$$

So, to prepare 50 g of 7.5% calcium chloride solution,

$$50. \text{ g solution} \times \frac{7.5 \text{ g calcium chloride}}{100. \text{ g solution}} = 3.8 \text{ g calcium chloride needed}$$

So 3.8 g of calcium chloride and 46.2 g of water would be needed.

Using the method described earlier for samples of pure liquids, determine the density of each of your sodium chloride solutions. Record the temperature of each solution while determining its density.

Using graph paper from the end of this manual, construct a graph of the **density** of your solutions versus the **percentage of NaCl** the solution contains. What sort of relationship exists between density and the composition of the solution?

Use a handbook of chemical data or the Internet to determine the true density of each of the solutions you prepared. Calculate the error in each of the densities you determined.

Name: _____ Section:_____

Lab Instructor: _____ Date:_____

EXPERIMENT 5

Density Determinations

Results/Observations

1. **Density of Solids**

    ID number of regular solid   _____

    Shape of regular solid   _____

    Dimensions of regular solid   _____

    Calculated volume of solid   _____

    Mass of regular solid   _____

    Density of regular solid   _____

    ID number of metal pellets   _____

    Mass of metal pellets taken   _____

    Initial water level   _____

    Final water level   _____

    Volume of metal pellets   _____

    Calculated density of metal pellets   _____

2. **Density of Pure Liquids**

    Mass of empty graduated cylinder   _____

    Mass of cylinder plus water   _____

    Volume of water   _____

    Density of water   _____

    Temperature of water   _____

    Handbook density of water   _____

    ID number of unknown liquid   _____

    Mass of cylinder plus liquid   _____

    Mass of liquid   _____

    Volume of unknown liquid   _____

    Density of unknown liquid   _____

3.   Density of Solutions

% NaCl	Density measured	Temperature	Handbook value	% error
5	_____	_____	_____	_____
10	_____	_____	_____	_____
15	_____	_____	_____	_____
20	_____	_____	_____	_____
25	_____	_____	_____	_____

Questions

1.   What error would be introduced into the determination of the density of the regularly shaped solid if the solid were hollow? Would the apparent volume of the solid be larger or smaller than the actual volume? Would the density calculated be too high or too low?

    _____

    _____

    _____

    _____

2.   What error would be introduced into the determination of the density of the irregularly shaped metal pellets if you had not stirred/shaken the pellets to remove adhering air bubbles? Would the density be too high or too low?

    _____

    _____

    _____

    _____

3.   Your data for the density of sodium chloride solutions should have produced a straight line when plotted. How could this plot be used to determine the density of *any* concentration of sodium chloride solution?

    _____

    _____

    _____

    _____

4.   What is the difference between the density of a solution and its specific gravity?

    _____

    _____

    _____

    _____

EXPERIMENT 6

Simple Distillation

Objective

Distillation is a common technique for the resolution of liquid mixtures. Simple distillation of an aqueous sodium chloride solution will be performed by the instructor as a demonstration to illustrate the method.

Introduction

Mixtures of liquids are most commonly separated by *distillation*. In general, distillation involves heating a liquid sample to its boiling point, then collecting, cooling and condensing the vapor that is produced into a separate container. For example, salt water can be desalinated by boiling off and condensing the water because salt is not volatile. But the separation of a mixture of liquids in which several of the components of the mixture are likely to be volatile is not so easy to effect.

If the components of the mixture differ considerably in their boiling points, it may be possible to separate the mixture simply by monitoring the temperature of the vapor produced as the mixture is heated. The components of a mixture will each boil in turn as the temperature is gradually raised; a sharp rise in the temperature of the vapor being distilled indicates that a new component of the mixture has begun to boil. By changing the receiving flask at this point, a separation will be accomplished.

For a mixture of liquids whose boiling points only differ by a few degrees, the mixture can be passed through a **fractionating column** as it is being heated. Fractionating columns are generally packed with glass beads or short lengths of glass tubing that provide a large amount of surface area to the liquid being boiled. In effect, a fractionating column permits a mixture to be redistilled repeatedly while in the column, allowing for better separation of the components of the mixture.

Safety Precautions	<ul><li>**Safety eyewear approved by your institution must be worn at all times while you are in the laboratory, whether or not you are working on an experiment.**</li><li>**Silver nitrate will stain the skin if spilled. Dispose of it as directed by the instructor.**</li><li>**Take precautions against burns when handling the distillation apparatus.**</li></ul>

Apparatus/Reagents Required

- 1% sodium chloride solution
- 0.1 *M* silver nitrate solution
- simple distillation set-up

Procedure

Record all observations and data directly on the report page in ink.

Simple distillation can be used when the components of a mixture have very different boiling points. In this experiment, a partial distillation of a solution of sodium chloride in water will be performed by the instructor (the distillation is not carried to completion in order to save time). This is an extreme example, since the boiling points of water and sodium chloride differ by over 1,000°C, but the technique will be clearly demonstrated by the experiment.

Your instructor has set up a simple distillation apparatus for you. (See Figure 6-1.) He or she will explain the various portions of the apparatus and will demonstrate the correct procedure for using the apparatus. The source of heat used for the distillation may be a simple burner flame, or an electrical heating device (heating mantle) may be provided. Generally, electrical heating elements are preferred for distillations if the substances being distilled are flammable.

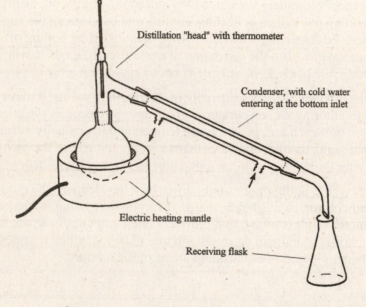

Distillation "head" with thermometer

Condenser, with cold water
entering at the bottom inlet

Electric heating mantle

Receiving flask

Figure 6-1.   Simple distillation apparatus
Cold water entering the lower inlet of the condenser causes the vapor being distilled to liquefy.

The instructor will obtain about 50 mL of 1% sodium chloride solution. He or she will place 1 mL of this solution in a small test tube, and will transfer the remainder of the solution to the distilling flask. A clean, dry beaker or flask will be placed under the mouth of the condenser of the distillation apparatus to collect the water as it distills.

The instructor will begin heating the sodium chloride solution and will continue distillation until approximately 20 mL of water has been collected. The instructor will transfer approximately 1 mL of the distilled water to a clean small test tube.

To demonstrate that the distilled water is now free of sodium chloride, the sample of original 1% sodium chloride solution that was reserved before the distillation, as well as the 1 mL sample of water that has been distilled, will be tested with a few drops of 0.1 M silver nitrate solution. Silver ion forms a precipitate of insoluble AgCl when added to a chloride ion solution. No precipitate should form in the water that has been distilled.

$$AgNO_3(aq) + NaCl(aq) \rightarrow AgCl(s) + NaNO_3(aq)$$

$$Ag^+(aq) + Cl^-(aq) \rightarrow AgCl(s)$$

EXPERIMENT 6

Simple Distillation

Pre-Laboratory Questions

1. Using your textbook or an encyclopedia of chemistry, describe the processes of *simple distillation* and *fractional distillation*. How do the methods differ from each other, and how are they similar?

    _____

    _____

    _____

    _____

    _____

    _____

    _____

    _____

    _____

    _____

    _____

    _____

2. Sketch the apparatus to be used for simple distillation.

3. Distillation has been suggested as a method for desalination of seawater for use as drinking water in desert countries. Use your textbook or an encyclopedia to discuss some of the problems associated with the desalination process, which has led to its being used only on a very limited basis.

    _____

    _____

    _____

    _____

    _____

    _____

    _____

    _____

    _____

    _____

4. Distillation is a common method applied to separate the components in mixtures of liquid substances. It can also be used to remove the solvent from the solute in a solution, as in this experiment. Use your textbook or an encyclopedia of chemistry to find three other methods of separating mixtures and describe briefly those methods.

    _____

    _____

    _____

    _____

    _____

    _____

    _____

    _____

    _____

    _____

    _____

    _____

Name: _____ Section: _____

Lab Instructor: _____ Date: _____

EXPERIMENT 6

Simple Distillation

Results/Observations

Sketch a representation of the simple distillation apparatus provided by your instructor. Identify each component of the apparatus and briefly describe its function.

Observation of distillation

_____

_____

_____

_____

_____

Silver nitrate test:

On original sample _____

On distilled sample _____

Questions

1. Write the net ionic reaction for the silver nitrate test used to detect chloride ion in this experiment.

   _____

2. The silver nitrate test should *not* have detected the presence of chloride ion in the distilled sample. Why? What type of substance is sodium chloride? Why would you not have expected sodium chloride to distill when the NaCl solution was heated? Hint: look up the properties of NaCl online or in a handbook of chemical data.

   _____

   _____

   _____

   _____

   _____

3. When a distillation is conducted, the mixture being distilled is *never* heated to complete dryness (even if this means some of the original mixture is lost). Suggest a reason why.

   _____

   _____

   _____

   _____

   _____

   _____

4. Some mixtures of substances cannot be separated by distillation (even a very careful fractional distillation) because the substances form what is called an *azeotrope* (or azeotropic mixture). For example, water and ethyl alcohol form an azeotrope that consists of 95% ethyl alcohol and 5% water. No matter how carefully this azeotrope is distilled, it is not possible to separate the two components. Use a chemical encyclopedia or dictionary to write the definition of an azeotrope.

   _____

   _____

   _____

   _____

   _____

   _____

   _____

EXPERIMENT 7

Properties of Some Representative Elements

Objective

In this experiment you will examine the properties of some of the more common representative elements.

Introduction

Even in the very earliest studies of chemistry, it became evident that certain elemental substances were very much like other substances in their physical and chemical properties. For example, the more common alkali metals (Na, K) were almost indistinguishable to early chemists. Both metals are similar in appearance and undergo reaction with the same reagents (giving formulas of the same stoichiometric ratio). As more and more elemental substances were separated, purified, and identified, more and more similarities between the new elements and previously known elements were detected. Chemists began to wonder why such similarities existed.

In the mid-1800s, Mendeleev and Meyer independently proposed that the properties of the known elements seemed to vary systematically when arranged in the order of their atomic masses; that is, when a list of the elements is made in order of increasing atomic mass, a given set of properties would repeat at regular intervals among the elements. Mendeleev generally receives most credit for the development of the periodic law because he suggested that some elements were missing from the table; that is, based on the idea that the properties of the elements should repeat at regular intervals, Mendeleev suggested that some elements that should have certain properties had not yet been discovered. Mendeleev went so far as to predict the properties of the as yet undiscovered elements, based on the properties of the already known elements. Mendeleev's periodic law received much acceptance when the missing elements were finally discovered and had the very properties that Mendeleev had predicted for them.

While Mendeleev arranged the elements in order of increasing atomic mass, the modern periodic table is arranged in order of increasing atomic number. At the time of Mendeleev's work, the structure of the atom had not yet been determined, but the relative masses of atoms had been determined. Generally, the arrangement of the elements by order of atomic mass is very similar to the arrangement by atomic number, with some notable exceptions. Some elements in Mendeleev's arrangement were out of order and did not have the properties expected. When the arrangement is made by atomic number, however, the properties of the elements do fall into a completely regular order, based on a repeating pattern of similar electronic structures.

Group 1 Elements

The Group 1 elements are commonly referred to as the **alkali metals**. These substances are among the most reactive of elements. They are never found in nature in the uncombined state, and are relatively difficult and expensive to produce and store. In particular, these elements are very easily oxidized by oxygen in the air; usually they are covered with a thin layer of oxide coating unless the metal has been freshly cut or cleaned. When obtained from a chemical supply house, these elements are usually stored under a layer of kerosene or other inert organic solvent to keep them from contact with the air.

The elements of Group 1 are very low in density (for example, they float on the surface of water while reacting with the water) and are soft enough to be easily cut with a knife or spatula. The only real reaction undergone by these elements involves the loss of their single valence electron to some other species:

$$M \rightarrow M^+ + e^-$$

In compounds, the elements of Group 1 are invariably found as the unipositive (1+) ion. The least reactive of these elements is lithium (the topmost member, with the valence electron held most tightly by the nuclear charge), whereas the most reactive is cesium (the bottommost member, in which the valence electron is very far from the nucleus). Cesium is so reactive that the metal may be ionized merely by shining light on it. This property of cesium has been utilized in certain types of photocells (in which the electrons produced by the ionization of cesium are channeled into a wire as electric current). The most common elements of this group, sodium and potassium, are found in great abundance in the combined state on the earth.

Group 2 Elements

The various elements of Group 2 are commonly referred to as the **alkaline earth elements**. The "earth" in this sense indicates that the elements are not as reactive as the alkali metals of Group 1. Beryllium, the topmost element of this group, is not very common. But some of the other members of the group, such as magnesium and calcium, are found in the combined state in relatively high abundance. As with the Group 1 elements, the relative reactivity of the Group 2 elements increases from top to bottom in the periodic table. For example, metallic magnesium can be kept for a relatively long time without developing an oxide coating (by reaction with air), whereas a freshly prepared sample of pure metallic calcium will develop a layer of oxide in a much shorter period. In compounds, the elements of Group 2 are usually found as dipositive (2+) ions (through loss of the two valence electrons).

$$M \rightarrow M^{2+} + 2e^-$$

Group 3 Elements

Whereas all the elements of Group 1 and Group 2 are metals, the first member of Group 3 (boron) is a nonmetal. The dividing line between metallic elements and nonmetallic elements forms the "stairstep" region indicated at the right-hand side of most periodic charts. Because boron is to the right of the stairstep, boron shows many properties that are characteristic of nonmetals. For example, boron is not a very good conductor of heat and electricity, whereas aluminum (the element beneath boron in the group) shows the much higher conductivities associated with metallic elements (aluminum is the conductor in most household wiring). Boron is generally found covalently bonded in its compounds, which further indicates its nonmetallic nature.

Aluminum is the only relatively common element from Group 3. In nature, aluminum is generally found as the oxide, and most metallic aluminum is produced by electrolysis of the molten oxide. This is a relatively difficult and expensive process, which is part of the reason the aluminum in cans was commonly recycled even before the present emphasis on preserving the environment. Aluminum is a relatively reactive element; however, this reactivity is sometimes masked under ordinary conditions. Aluminum used for making cooking utensils or cans becomes very quickly coated with a thin layer of aluminum oxide, Al_2O_3. This oxide layer serves as a protective coating, preventing further oxidation of the aluminum metal underneath. (This tendency explains why aluminum pans are seldom as shiny as copper or stainless steel pans.)

Group 4 Elements

Group 4 contains some of the most common, useful and important elements known. The first member of the group, carbon, is a nonmetal. The second and third members, silicon and germanium, show both some metallic and some nonmetallic properties (and because of this in-between nature, these elements are

54

referred to as **metalloids** or **semimetals**). The last two members of the group, tin and lead, show mostly metallic properties.

The element carbon forms the framework for life. Virtually all biological molecules are, in fact, carbon compounds. Carbon is unique among the elements in that it is able to form long chains of many hundreds or even thousands of similar carbon atoms. Pure elemental carbon itself is usually obtained in either of two forms, graphite or diamond, each of which has a very different structure. Graphite contains flat, two-dimensional layers of covalently bonded carbon atoms, with each layer effectively being a molecule of graphite, independent of the other layers present in the sample. The layers of carbon atoms in graphite are able to move relative to each other. This property makes graphite slippery and useful as a lubricant for machinery and locks. Diamond contains three-dimensionally extended, covalently-bonded networks of carbon atoms, meaning an entire diamond is, effectively, a single molecule. Because all the atoms present in diamond are covalently bonded to each other in three dimensions, diamond is relatively stable (diamond will burn in oxygen, however, producing carbon dioxide and water). Because of the extensive bonding among the carbon atoms, diamond is the hardest substance known. Natural diamonds used as gemstones are produced deep in the earth under high pressures over eons, but synthetic diamonds are produced in the laboratory by compressing graphite to several thousand atmospheres of pressure. Because of its hardness, diamond is used as an abrasive in industry.

Silicon and germanium are used in the semiconductor industry. Because these elements have some of the properties of nonmetals (brittleness, hardness), as well as some of the properties of metals (slight electrical conductivity), they have been used extensively in transistors.

The last two elements of Group 4, tin and lead, show mostly metallic properties (shiny luster, conductivity, malleability) and frequently form ionic compounds. However, covalently bonded compounds of these metals are not rare. For example, leaded gasoline contains tetraethyllead, which is a covalently-bonded, organic lead compound. Tin has been used in the past to make cooking utensils and is used extensively as a liner for steel cans ("tin" cans) because it is less likely to be oxidized than iron and is also less likely to impart a bad taste to foods. Whereas tin is nonpoisonous and is used in the food industry, lead compounds are very dangerous poisons. Formerly, pigments for white paint were made using lead oxide. Lead compounds are remarkably sweet-tasting, and many children have been seriously poisoned by eating flecks of peeling lead paints.

Group 5 Elements

The first few elements of Group 5 are nonmetals, with nitrogen and phosphorus being the two most abundant elements of the group. The earth's atmosphere is nearly 80% N_2 gas by volume. This large percentage of nitrogen is believed to be a remnant of the composition of the earth's atmosphere at the time the solar system formed. Astronomers have determined that several of the planets of the solar system contain large quantities of ammonia, NH_3, in their atmospheres. It is known that ammonia actually comes to equilibrium with its constituent elements, that is:

$$2NH_3 \rightarrow N_2 + 3H_2$$

Astronomers theorize that the atmosphere of the early earth was also mostly ammonia, but that the lighter hydrogen gas produced by the equilibrium has escaped the earth's relatively weak gravitational field over the eons, leaving the atmosphere with a large concentration of nitrogen gas. Nitrogen is also an important constituent of many of the molecules synthesized and used by living cells (amino acids, for example). Nitrogen is one of the most important nutrients for plants, and certain bacteria in the soil are able to capture nitrogen from the atmosphere for use by plants (nitrogen fixation).

Phosphorus, the second member of Group 5, is a typical solid nonmetal. Three allotropic forms of elemental phosphorus exist: red, white and black phosphorus. White phosphorus is the most interesting of the three forms, and is the form that is produced when phosphorus vapor is condensed. White phosphorus is very unstable toward oxidation. Below about 35°C, white phosphorus reacts with oxygen of the air, emitting the energy of the oxidation as light rather than heat (phosphorescence); above 35°C, white phosphorus will spontaneously burst into flame. (For this reason, white phosphorus is stored under water to keep it from contact with air.)

Whereas elemental nitrogen consists of diatomic molecules (N_2), white phosphorus contains tetrahedral-shaped P_4 molecules. When white phosphorus is heated to high temperatures in the absence of air, it is converted to the red allotrope. The red allotrope is much less subject to oxidation and is the form usually available in the laboratory. The lower reactivity is due to the fact that the phosphorus atoms have polymerized into one large molecule, which is far less subject to attack by oxygen atoms than the individual P_4 molecules of the white allotrope.

Phosphorus is essential to life, being a component of many biological molecules, most notably the phosphate compounds used by the cell for storing and transferring energy (AMP, ADP and ATP). Phosphorus also forms part of the structural framework of the body; teeth and bones are composed of complex calcium phosphate compounds.

The other members of Group 5 are far less common than are nitrogen and phosphorus. For example, arsenic (As) is known primarily for its severe toxicity and has been used in various rodent and weed killers (though As has been supplanted by other agents for most of these purposes).

Group 6 Elements

The most abundant elements of Group 6 (sometimes referred to as the *chalcogens*) are oxygen and sulfur, which, along with selenium, are nonmetallic. Elemental oxygen makes up about 20% of the atmosphere by volume and is vital not only to living creatures but to many common chemical reactions. Most of the oxygen in the atmosphere is produced by green plants, especially plankton in the oceans of the earth. On a microscopic basis, oxygen is used in living cells for the oxidation of carbohydrates and other nutrients. Oxygen is needed also for the oxidation of petroleum-based fuels for heating and lighting purposes and for myriad other important industrial processes. Oxygen has two allotropic forms: the normal elemental form of oxygen is O_2 (dioxygen), whereas a less stable allotrope, O_3 (ozone), is produced in the atmosphere by high-energy electrical discharges (e.g., lightning).

Sulfur is obtained from the earth in a nearly pure state and needs little or no refining before use. In certain areas of the earth, particularly where there has been volcanic action, there are vast underground deposits of relatively pure sulfur. High-pressure steam is pumped into such deposits, liquefying the sulfur and allowing it to be pumped to the surface of the earth for collection (Frasch process). The main use of sulfur is in the manufacture of sulfuric acid, which is the industrial chemical produced in the largest amount. When sulfur is burned in air, it is converted to sulfur dioxide, notable for its choking, irritating odor (similar to a freshly struck match). If sulfur dioxide is further oxidized with a catalyst, it is converted to sulfur trioxide, the anhydride of sulfuric acid. Many fossil fuels contain sulfur, and on oxidation, SO_2 and SO_3 may be released into the atmosphere unless precautions are taken: this release has resulted in the acid rain problem now being experienced in many parts of the world.

Sulfur has several allotropic forms. At room temperature and pressure, the normal allotrope is orthorhombic sulfur, which consists of S_8 cyclic molecules. At higher temperatures, a slight rearrangement of the shape of these rings occurs, producing a solid allotropic form of sulfur called monoclinic sulfur. This substance has a different crystalline structure than the orthorhombic form. (Monoclinic sulfur slowly changes into orthorhombic sulfur at room temperature.) The most interesting allotrope of sulfur, however, is produced when boiling sulfur is rapidly lowered in temperature. This technique produces a plastic (amorphous) allotrope. The properties of plastic sulfur are very different from those of orthorhombic or monoclinic sulfur. Whereas plastic sulfur is soft and pliable (able to be stretched and pulled into strings), orthorhombic and monoclinic sulfur are both hard, brittle solids. Sulfur is insoluble in water, but the orthorhombic form is fairly soluble in the nonpolar solvent carbon disulfide, CS_2.

The most common oxygen compound is, of course, water, H_2O. The analogous sulfur compound, hydrogen sulfide, H_2S, is a noxious-smelling gas. (The gas smells like rotten eggs; eggs contain a protein involving sulfur that releases hydrogen sulfide as the egg spoils.)

Selenium, which occurs in Group 6 just beneath sulfur, is the only other member of the group that occurs in any abundance. Selenium compounds are generally very toxic. Selenium is used in certain prescription dandruff shampoos and treatments.

Group 7 Elements

The elements of Group 7 are more commonly referred to as the *halogens* ("salt-formers"). The elementary substances are very reactive; these elements are, therefore, most commonly found in the combined state, as negative halide ions. Elemental fluorine is the most reactive nonmetal and is seldom encountered in the laboratory because of problems of toxicity and handling of the gaseous element. Most compounds of fluorine are very toxic, although tin(II) fluoride is used in toothpastes and mouthwashes as a decay preventative: fluoride ion can replace hydroxide ion in the complex mixture that constitutes teeth, resulting in a much harder, nonreactive surface.

Hydrogen fluoride is very different from the other hydrogen halides, since aqueous solutions of HF are weakly acidic, rather than strongly acidic (HCl, HBr). Hydrogen fluoride is able to attack and dissolve glass and is always stored in plastic bottles.

Gaseous elemental chlorine is somewhat less reactive than elemental fluorine and is used commercially in solution as a disinfectant. Chlorine will oxidize biological molecules present in bacteria (thereby killing the bacteria), while it is reduced itself to chloride ion (which is nontoxic). For example, drinking water is usually chlorinated, either with gaseous elemental chlorine or with some compound of chlorine that is capable of releasing elemental chlorine when needed. Gaseous hydrogen chloride, when dissolved in water, forms the solution known as hydrochloric acid, which is one of the most commonly used acids.

Elemental bromine is a strikingly dark red liquid at room temperature. Elemental bromine is used as a common laboratory test for the presence of double bonds in carbon compounds. Br_2 reacts with the double bond, and its red color disappears, thereby serving as an indicator of reaction.

Elemental iodine is a dark gray solid at room temperature but undergoes sublimation when heated slightly, producing an intensely purple vapor. Elemental iodine is the least reactive of the halogens. Iodine is very important in human metabolism, being a component of various hormones produced by the thyroid gland. Since many people's diet does not contain sufficient iodine from natural sources, commercial table salt usually has a small amount of sodium iodide added as a supplement ("iodized" salt). Iodine is also used sometimes as a topical antiseptic, since it is a weak oxidizing agent that is capable of destroying bacteria. Iodine for this purpose (called "tincture" of iodine) is usually sold in alcohol solution.

Safety Precautions	<ul><li>Safety eyewear approved by your institution must be worn at all times while you are in the laboratory, whether or not you are working on an experiment.</li><li>The reactions of lithium, sodium and potassium with water are very dangerous. Your instructor will demonstrate these reactions for you. *Do not attempt these reactions yourself.*</li><li>Hydrogen gas is extremely flammable and forms explosive mixtures with air. Small amounts of hydrogen will be generated in some of the student reactions. Avoid flames and exercise caution when dealing with hydrogen.</li><li>The flames produced by the burning of magnesium and calcium are intensely bright and can damage the eyes. Do not look directly at the flame while these substances are burning.</li><li>Oxides of nitrogen are toxic and extremely irritating to the respiratory system. Generate these gases only in the fume exhaust hood.</li><li>Hydrochloric and nitric acids will burn skin, eyes and clothing. Wash immediately if the acids are spilled, and inform the instructor.</li><li>Vapors of methylene chloride and the halogens are toxic. Keep these substances in the fume exhaust hood at all times.</li><li>Dry ice is extremely cold, and should only be handled with tongs or thermal gloves.</li><li>Hydrogen sulfide and sulfur dioxide gases are toxic and have noxious odors. Confine these gases to the fume exhaust hood.</li><li>Salts of the Group 1 and 2 metals used for the flame tests may be toxic. Wash after handling these compounds.</li><li>Dispose of all materials as directed by the instructor.</li></ul>

Apparatus/Reagents Required

- lithium (instructor only)
- sodium (instructor only)
- potassium (instructor only)
- universal indicator solution
- calcium
- magnesium ribbon

- samples of solid LiCl, NaCl, KCl, $CaCl_2$, $BaCl_2$ and $SrCl_2$
- flame test wires
- 6 M hydrochloric acid
- boric acid
- aluminum oxide
- calcium oxide
- dry ice
- nichrome wire
- sodium peroxide
- chlorine water
- bromine water
- iodine water
- methylene chloride
- 0.1 M NaCl
- 0.1 M NaBr
- 0.1 M NaI
- ferrous sulfide
- sulfur
- solid iodine
- ice

Procedure

Record all data and observations directly on the report pages in ink.

1. Some Properties of the Alkali and Alkaline Earth Elements

The reactive metals of Group 1 and Group 2 react with water to liberate elemental hydrogen, leaving a solution of the strongly basic metal hydroxide:

$$2M(s) + 2H_2O \rightarrow 2M^+(aq) + 2OH^-(aq) + H_2(g)$$

$$M(s) + 2H_2O \rightarrow M^{2+}(aq) + 2OH^-(aq) + H_2(g)$$

Generally the reactivity of the metals increases going from the top of the group toward the bottom of the group. The valence electrons of the metal are less tightly held by the nucleus for the elements towards the bottom of the group.

a. Lithium, Sodium and Potassium (Instructor Demonstration)

The reaction of the Group 1 metals with water is very dangerous and will be performed by your instructor as *demonstrations*. **Do not attempt these reactions yourself.**

In the exhaust hood, behind a safety shield, the instructor will drop small pellets of lithium, sodium, and potassium into separate beakers containing a small amount of water. Compare the speed and vigor of the reactions.

When the reactions have subsided, obtain small portions of the water from the beakers in which the reactions were conducted in separate clean test tubes. Add two or three drops of universal indicator to each test tube. Refer to the color chart provided with the indicator to determine the pH of the solution. Write equations for the reactions of the metals with water. Why are the solutions produced basic?

b. Reactions of Magnesium and Calcium (Student Procedure)

Repeat the procedure that you have seen in the demonstration, using pieces of the Group 2 metals magnesium and calcium in place of lithium, sodium and potassium. Work in the exhaust hood, with the safety shield pulled down. Account for differences in reactivity between these two Group 2 metals. Test the pH of the water with which the metals were reacted.

Add a short strip of magnesium ribbon to a few milliliters of 6 M HCl in a small beaker. Why does magnesium react with acid but not with water?

c. Flame Tests

Obtain a 6- to 8-inch length of nichrome wire for use as a flame test wire. Form one end of the wire into a small loop that is no more than a few millimeters in diameter.

Obtain about 10 mL of 6 M HCl in a small beaker, and immerse the loop end of the wire for 2–3 minutes to clean the wire.

Ignite a Bunsen burner flame, and heat the loop end of the wire in the flame until it no longer imparts a color to the flame.

Obtain small samples of LiCl, NaCl, KCl, CaCl$_2$, BaCl$_2$ and SrCl$_2$. Dip the loop of the flame test wire into one of the salts, and then into the oxidizing portion of the burner flame. Record the color imparted to the flame by the salt.

Clean the wire in 6 M HCl, heat in the flame until no color is imparted, and repeat the flame test with the other salts. The colors imparted to the flame by the metal ions are so intense and so characteristic that they are frequently used as a test for the presence of these elements in a sample.

2. Oxides/Sulfides of Some Elements

Oxygen compounds of most elements are known, and are generally referred to as **oxides**. However, there are great differences between the oxides of metallic elements and those of nonmetals. The oxides of metallic elements are ionic in nature: they contain the oxide ion, O^{2-}. Ionic metallic oxides form basic solutions when dissolved in water

$$O^{2-} + H_2O \rightarrow 2OH^-$$

Oxides of nonmetals are generally covalently bonded and form acidic solutions when dissolved in water. For example, the oxide of phosphorus, P_2O_5, is acidic in water since it produces phosphoric acid when dissolved

$$P_2O_5(s) + 3H_2O(l) \rightarrow 2H_3PO_4(aq)$$

Sulfur compounds of both metallic and nonmetallic substances are known. Sulfur compounds of nonmetallic elements are generally covalently bonded, whereas with metallic ions, sulfur is usually present as the sulfide ion, S^{2-}. Most sulfur compounds have characteristically unpleasant odors, or decompose into compounds that have the characteristic odor.

a. Metallic Oxides

Place a small amount of distilled water in a beaker and add two or three drops of universal indicator. Refer to the color chart provided with the indicator, and record the pH of the water.

Obtain a piece of magnesium ribbon about 1 inch in length. Hold the magnesium ribbon with tongs above the beaker of water and ignite the metal with a burner flame. Do *not* look directly at the flame produced by burning magnesium. It is intensely bright and damaging to the eyes. When the flame has expired, stir the liquid in the beaker for several minutes and record the pH of the solution. Write an equation for the reactions that have taken place.

Place approximately 5 to 6 mL of distilled water in each of three test tubes, and add two or three drops of universal indicator to each. Add a very small quantity of sodium (per)oxide to one test tube, calcium oxide (lime) to a second test tube, and aluminum oxide (alumina) to the third test tube. Stir each test tube with a clean glass rod, record the pH of the solution, and write an equation to account for the pH.

b. Nonmetallic Oxides

Place approximately 5 to 6 mL of water into each of two test tubes, and add two or three drops of universal indicator to each. In one test tube, dissolve a small amount of boric acid. Using tongs or thermal gloves to protect your hands from the cold, add a small chunk of dry ice (solid carbon dioxide) to the second test tube. Stir the test tubes with a clean glass rod, record the pH, and write an equation to account for the pH.

c. Sulfur Compounds

Work with sulfur and sulfur compounds in the fume exhaust hood.

Have ready a beaker of cold water. Ignite a *tiny* amount of powdered sulfur on the tip of a spatula in the burner flame and *cautiously* note the odor. Hold the burning sulfur approximately 1 cm above the surface of the water in the beaker for a few seconds, then extinguish the sulfur in the cold water to prevent too much SO_2 from getting into the air. Determine the pH of the water in which you extinguished the sulfur. Why is it acidic?

Obtain a *tiny* portion of iron(II) sulfide in a test tube, and add 2 drops of dilute hydrochloric acid. *Cautiously* note the odor of the hydrogen sulfide generated. Transfer the test tube to the exhaust hood to dispose of the hydrogen sulfide produced.

3. Some Properties of the Halogen Family (Group 7)

All of the members of Group 7 are nonmetallic. The halogen elements tend to gain electrons in their reactions, and the attraction for electrons is stronger if the electrons are closer to the nucleus of the atom. The activity of the halogen elements therefore decreases from top to bottom in Group 7 of the periodic table. As a result of this, elemental chlorine is able to replace bromide and iodide ion from compounds:

$$Cl_2(g) + 2NaBr(aq) \rightarrow Br_2(l) + 2NaCl(aq)$$

$$Cl_2(g) + 2NaI(aq) \rightarrow I_2(s) + 2NaCl(aq)$$

Similarly, elemental bromine is able to replace iodide ion from compounds.

a. Identification of the Elemental Halogens by Color

The elemental halogens are nonpolar and are not very soluble in water. Solutions of the halogens in water are not very brightly colored. However, if an aqueous solution of an elemental halogen is shaken with a nonpolar solvent, the halogen is preferentially extracted into the nonpolar solvent, and imparts a characteristic bright color to the nonpolar solvent. Since the nonpolar solvent is not miscible with water, the halogen color is evident as a separate, colored layer.

Obtain about 1 mL of chlorine water in a test tube. Note the color. In the exhaust hood, add 10 drops of methylene chloride to the test tube. Stopper and shake.

Allow the solvent layers to separate and note the color of the lower (methylene chloride) layer. Elemental chlorine is not very intensely colored and imparts only a pale yellow/green color.

Repeat the procedure using 1 mL of bromine water in place of the chlorine water. Bromine imparts a bright red color to the lower layer.

Repeat the procedure using 1 mL of iodine water. Iodine imparts a purple color.

b. Relative Reactivity of the Halogens

Add 10 drops of chlorine water to each of two semi-micro test tubes. Add about 10 drops of 0.1 M NaBr to one test tube and 10 drops of 0.1 M NaI to the second tube.

In the exhaust hood, add 10 drops of methylene chloride to each test tube. Stopper, shake and allow the solvents to separate. Record the colors of the lower layers and identify which elemental halogens have been produced.

Add 10 drops of bromine water to each of two semi-micro test tubes. Add about 10 drops of 0.1 M NaI to one test tube and 10 drops of 0.1 M NaCl to the other test tube.

In the exhaust hood, add 10 drops of methylene chloride to each test tube. Stopper, shake, and allow the layers to separate. Record the colors of the lower layers. Indicate where a reaction has taken place.

Add 10 drops of iodine water to each of two semi-micro test tubes. Add about 10 drops of 0.1 M NaBr to one test tube and 10 drops of 0.1 M NaCl to the other test tube.

In the exhaust hood, add 10 drops of methylene chloride to each test tube. Stopper, shake, and allow the layers to separate. Record the colors of the lower layer. Did any reaction take place?

c. Sublimation of Iodine

Using forceps, place a few crystals of elemental iodine in a small beaker. Set the beaker on a ring stand in the exhaust hood, and cover the beaker with a watch glass containing some ice. Heat the iodine crystals in the beaker with a small flame for 2–3 minutes, and watch the sublimation of the iodine. Extinguish the flame, and examine the iodine crystals that have formed on the lower surface of the watch glass.

EXPERIMENT 7

Properties of Some Representative Elements

Pre-Laboratory Questions

1. For each of the following elements, list the atomic number, average atomic molar mass, in which group (vertical column) of the periodic table the element can be found, and also in which period (horizontal row) the element is located.

Element Symbol	Atomic Number	Molar Mass	Group	Period
Mg	_____	_____	_____	_____
N	_____	_____	_____	_____
F	_____	_____	_____	_____
K	_____	_____	_____	_____
Si	_____	_____	_____	_____
Ra	_____	_____	_____	_____
Xe	_____	_____	_____	_____
Cs	_____	_____	_____	_____

2. Why do members of the same vertical group of the periodic chart tend to show similar chemical and physical properties?

_____

_____

_____

_____

_____

_____

_____

_____

_____

_____

3.   List the symbols of the first three elements (from top to bottom on the Periodic Table) in each of the
     following families.

     alkali metals   _____   _____   _____

     halogens   _____   _____   _____

     Noble gases   _____   _____   _____

     alkaline earths   _____   _____   _____

     chalcogens   _____   _____   _____

4.   Reactions of sodium, lithium and potassium will be demonstrated by your instructor in this
     experiment. Why are students not allowed to handle these elements?

     _____

     _____

     _____

     _____

     _____

     _____

     _____

     _____

     _____

     _____

5.   Reactions involving sulfur dioxide, hydrogen sulfide, the halogens and methylene chloride will be
     performed in the fume exhaust hood. Why?

     _____

     _____

     _____

     _____

     _____

     _____

     _____

     _____

     _____

Name:_____ Section:_____

Lab Instructor: _____ Date:_____

EXPERIMENT 7

Properties of Some Representative Elements

Results/Observations

1. Reactions of Alkali/Alkaline Earth Metals

    a. Reaction with water

Metal	Evidence of reaction	Relative vigor of reaction	Color of indicator
Li	_____	_____	_____
Na	_____	_____	_____
K	_____	_____	_____
Mg	_____	_____	_____
Ca	_____	_____	_____

Balanced equations for reactions that occurred:

_____

_____

_____

_____

_____

    b. Observation of reaction of Mg with acid:

_____

Balanced equation for reaction:

_____

c.   Flame tests

Metal	Color observed
_____	_____
_____	_____
_____	_____
_____	_____
_____	_____
_____	_____

2.   Oxides/Sulfides of Some Elements

Substance	Results/equations	Universal indicator test
magnesium oxide	_____	_____
sodium oxide	_____	_____
calcium oxide	_____	_____
aluminum oxide	_____	_____
boric acid	_____	_____
dry ice	_____	_____
sulfur dioxide	_____	_____
FeS + HCl	_____	_____

3.   Some Properties of the Halogen Family

a.   Identification of the elemental halogens

Halogen	Color of water solution	Color of CH_2Cl_2 extract
Cl_2	_____	_____
Br_2	_____	_____
I_2	_____	_____

b.   Relative reactivity of the halogens

Elemental halogen	Halide ion	Color of lower layer	Which halogen is in lower layer?	Did a reaction take place?
Cl_2	Br^-	_____	_____	_____
Cl_2	I^-	_____	_____	_____
Br_2	I^-	_____	_____	_____
Br_2	Cl^-	_____	_____	_____
I_2	Br^-	_____	_____	_____
I_2	Cl^-	_____	_____	_____

Equations for reactions of the halogens that occurred:

_____

_____

_____

_____

_____

c.   Sublimation of iodine observation:

_____

_____

_____

_____

_____

Questions

1.   In comparing the vigor of the reactions of lithium, sodium and potassium with cold water, does there appear to be a correlation between the location of these elements in the periodic table and the vigor of reaction?

_____

_____

_____

_____

_____

_____

_____

2.   In general, oxides of the metallic elements are basic when dissolved in water, whereas oxides of the non-metallic elements behave as acids when dissolved in water. Using the elements you have studied in this experiment, give two examples for metallic oxides and two examples for non-metallic oxides demonstrating this general observation. Use chemical equations to make your point.

_____

_____

_____

_____

_____

_____

3. Elemental fluorine, F_2, was not used in the halogens experiment because of difficulties in handling it, and elemental astatine, At_2, was not used because of its radioactivity and short half-life. How would you expect the reactivity of F_2 and At_2 to compare with those of the elemental halogens tested in this experiment? Would F_2 be more or less reactive than Cl_2? Why? Would At_2 be able to displace iodine from a solution of potassium iodide? Why?

_____

_____

_____

_____

_____

_____

_____

4. When you performed flame tests on the chloride salts of several metallic elements, you undoubtedly noticed that each element imparted a different color to the flame. These colors are *characteristic* for these elements. Use your textbook or a chemical encyclopedia to explain why these metals always impart their characteristic colors to a flame.

_____

_____

_____

_____

_____

_____

_____

5. In this experiment, you studied the sublimation of elemental iodine: under atmospheric pressure, iodine passes directly from the solid state to the vapor state without ever forming a liquid. Give two additional examples of elements or compounds that undergo sublimation. **Hint**: there was another substance in this experiment that undergoes sublimation.

_____

_____

_____

_____

_____

_____

_____

EXPERIMENT 8

Thin-Layer Chromatography

Objective

The field of **separation science** is one of the most important in chemistry today. The particular branch of chemistry called **analytical chemistry** is concerned with the separation of mixtures and the analysis of the amount of each component in the mixture. In this experiment, you will perform the separation of a mixture of colored dyes by the technique of thin-layer chromatography.

Introduction

The word **chromatography** means "color writing." The name was chosen at the beginning of the 20th century when the method was first used to separate colored components from plant leaves. Chromatography in its various forms is perhaps the most important known method for the chemical analysis of mixtures.

The earliest form of chromatography – paper chromatography – was performed using ordinary filter paper, which consists primarily of the polymeric carbohydrate cellulose, as the medium upon which the mixture to be separated is applied. The more modern technique of **thin-layer chromatography** (universally abbreviated as **TLC**) uses a thin coating of aluminum oxide (alumina) or silicagel on a glass microscope slide or plastic sheet, to which the mixture to be resolved is applied.

In TLC, a single drop or spot of the unknown mixture to be analyzed is applied about half an inch from the end of a TLC slide. The TLC slide is then placed in a shallow layer of solvent mixture in a jar or beaker. Since the coating of the TLC slide is permeable to liquids, the solvent begins rising through the coating by capillary action.

As the solvent rises to the level at which the spot of mixture was applied, various effects can occur, depending on the constituents of the spot. Those components of the spot that are completely soluble in the solvent will be swept along with the solvent front as it continues to rise. The components that are not at all soluble in the solvent will be left behind at the original location of the spot. Most components of the unknown spot mixture will take an intermediate approach as the solvent front passes. Components in the spot that are somewhat soluble in the solvent will be swept along by the solvent front, but to different extents, reflecting their specific solubilities. By this means, the original spot of mixture is spread out into a series of spots or bands, with each spot representing one single component of the original mixture.

The separation of a mixture by chromatography is not solely a function of the solubilities of the components in the solvent used, however. The TLC slide coating used in chromatography is not inert, but consists of molecules that may interact with the molecules of the components of the mixture being separated. Each component of the mixture is likely to have a different extent of interaction with the slide coating. This differing extent of interaction between the components of a mixture and the molecules of the coating on the TLC slide forms an equally important basis for the separation. The coating of the TLC slide adsorbs molecules on its surface to differing extents, depending on the structure and properties of the molecules involved.

To place a TLC separation on a quantitative basis, a mathematical function called the **retention factor**, R_f, is defined:

$$R_f = \frac{\text{distance travelled by spot}}{\text{distance travelled by solvent}}$$

The retention factor depends on what solvent is used for the separation and on the specific composition of the slide coating used for a particular analysis. Because the retention factors for particular components of a mixture may vary if an analysis is repeated under different conditions, a known sample is generally analyzed at the same time as an unknown mixture on the same TLC slide. If the unknown mixture produces spots having the same R_f values as spots from the known sample, then an identification of the unknown components has been achieved.

Thin-layer chromatography is only one example of the many different chromatographic methods available. Mixtures of volatile liquids are commonly separated by a method called **gas chromatography**. In this method, a mixture of liquids is vaporized and passed through a long tube of solid adsorbent material (coated with an appropriate liquid), via the action of a carrier gas (usually an inert gas such as helium). As with TLC, the components of the mixture will have different solubilities in the liquid coating and different attractions for the solid adsorbent material. Separation of the components of the mixture occurs as the mixture progresses through the tube. The individual components of the mixture exit the tube one by one and are usually detected by electronic means.

In this experiment, you will perform a thin-layer chromatographic analysis of a mixture of the dyes bromcresol green, methyl red and xylenol orange. These dyes have been chosen because they have significantly different retention factors, and a nearly complete separation should be possible in the appropriate solvent system. You will also investigate the effect of the solvent on TLC analyses by attempting the separation in several different solvent systems.

In real practice, thin-layer chromatography has several uses. When a new compound is synthesized, for example, a TLC of the new compound is routinely done to make certain that the new compound is pure (a completely pure compound should only give a single TLC spot; impurities would result in additional spots). TLC is also used to separate the components of natural mixtures isolated from biological systems: for example, the various pigments in plants can be separated by TLC of an extract made by boiling the plant leaves in a solvent. Once the components of a mixture have been separated by TLC, it is even possible to isolate small quantities of each component by scraping its spot from the TLC slide and redissolving the spot in some suitable solvent.

Safety Precautions	• **Safety eyewear approved by your institution must be worn at all times while you are in the laboratory, whether or not you are working on an experiment.** • **The organic indicator dyes used in this experiment will stain skin and clothing if spilled. Many such dyes are toxic or mutagenic.** • **The solvents used for the chromatographic separation are *highly flammable* and their vapors are toxic. *No flames should be burning in the room while these solvents are in use. Use the solvents only in the fume exhaust hood.* Dispose of the solvent mixtures as directed by the instructor: do *not* pour down the drain.**

Apparatus/Reagents Required

- Bakerflex® plastic TLC slides (1 × 4 inch)

- surgical gloves

- ruler

- pencil

- plastic wrap or Parafilm®

- 10 microliter micropipets

- ethanol solutions of the indicator dyes methyl red, xylenol orange bromcresol green (or other dyes as provided by your instructor)

- acetone

- ethyl acetate

- hexane

- ethanol

Procedure

Clean and dry six 400-mL beakers to be used as the chambers for the chromatography. Obtain several squares of plastic wrap or Parafilm® to be used as covers for the beakers.

The chromatographic separation will be attempted in several solvent mixtures to investigate which solvent mixture gives the most complete resolution of the three dyes. A total of only 10 to 15 mL of each solvent mixture is necessary. Prepare mixtures of the solvents below, in the proportions indicated by volume, and transfer each to a separate 400-mL beaker. Cover the beakers after adding the solvent mixture, and label the beakers with the identity of the mixture each contains.

acetone 60% / hexane 40%

ethyl acetate 60% / hexane 40%

acetone 50% / ethyl acetate 50%

acetone 50% / ethanol 50%

ethyl acetate 50% / ethanol 50%

hexane 50% / ethanol 50%

Wearing plastic surgical gloves to avoid any oils from the fingers, prepare six plastic TLC slides by marking *lightly* with pencil (not ink) a line across both the top and bottom of the slide. Do not mark the line too deeply or you will remove the coating of the slide. See Figure 8-1 on the following page.

On one of the lines you have drawn on each slide, mark four small pencil dots (to represent where the spots are to be applied).

Above the other line on each slide, mark the following letters: *R* (methyl red), *X* (xylenol orange), *G* (bromcresol green) and *M* (mixture). (See Figure 8-1 on the following page.) If your instructor provided dyes other than the three mentioned above, rather than using *R*, *X*, and *G* to indicate the dyes on the TLC slide, you may use some other abbreviations that will help you to distinguish the three dyes.

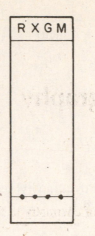

Figure 8-1. Plastic TLC slide with spots of the three dyes and the mixture applied
Keep the spots you apply as small as possible.

Obtain small samples of the ethanolic solutions of the three dyes (methyl red, xylenol orange bromcresol green). Also obtain several micropipets: use a separate micropipet for each dye, and be careful not to mix up the pipets during the subsequent application of the dyes.

Apply a single small droplet of the appropriate dye to its pencil spot on each of the TLC slides you have prepared (wipe the outside of the micropipet if necessary before applying the drop to remove any excess dye solution). Keep the spots of dye as small as possible.

Apply one droplet of each dye to the spot labeled *M* (mixture) on each slide, being sure to allow each previous spot to dry before applying the next dye. Allow the spots on the TLC slides to dry before proceeding.

Position the beakers containing the various solvent systems in a location where you will be able to leave them undisturbed for several minutes.

Gently lower one of the TLC slides, spots downward, into one of the solvent systems. Be careful not to wet the spots, or to slosh the solvent in the beaker; do not move or otherwise disturb the beaker after adding the TLC slide. Carefully cover the beaker with plastic wrap.

Allow the solvent to rise on the TLC slide until it reaches the upper pencil line (this will not take very long).

When the solvent has risen to the upper pencil mark, remove the TLC slide and quickly mark the exact solvent front before it evaporates. Mark the TLC slide with the identity of the solvent system used for development. Set the TLC slide aside to dry completely.

Repeat the process using the additional TLC slides and solvent systems. Be certain to mark each slide with the solvent system used.

Using a ruler, determine R_f for each dye in each solvent system and record. Which solvent system led to the most complete resolution of the dye mixture? If no mixture gave a complete resolution, your instructor may suggest other solvents for you to try, or other proportions of the solvents already used. Save your TLC slides and staple them to the lab report page for this experiment.

EXPERIMENT 9

Calorimetry

Objective

A simple "coffee cup" calorimeter will be used to measure the quantity of heat that flows in two strong acid/strong base neutralization reactions.

Introduction

Chemical and physical changes are always accompanied by a change in energy. Most commonly, this energy change is observed as a flow of heat energy either into or out of the system under study. Heat flows are measured in an instrument called a **calorimeter**. There are specific types of calorimeters for specific reactions, but all calorimeters contain the same basic components. They are insulated to prevent loss or gain of heat energy between the calorimeter and its surroundings. For example, the simple calorimeter you will use in this experiment is made of a heat-insulating plastic foam material. Calorimeters contain a heat sink that can absorb or provide the energy for the process under study. The most common material used as a heat sink for calorimeters is water because of its ready availability and large heat capacity. Calorimeters also must contain some device for the measurement of temperature, because it is from the temperature change of the calorimeter and its contents that the magnitude of the heat flow is calculated. Your simple calorimeter will use an ordinary thermometer for this purpose.

To determine the heat flow for a process, the calorimeter typically is filled with a weighed amount of water. The process that releases or absorbs heat is then performed within the calorimeter, and the temperature of the water in the calorimeter is monitored. From the mass of water in the calorimeter, and from the temperature change of the water, the quantity of heat transferred by the process can be determined.

When a sample of *any* substance changes in temperature, the quantity of heat, Q, involved in the temperature change is given by:

$$Q = mC\Delta T$$

In this equation, m is the mass of the substance, ΔT is the temperature change, and C is a quantity called the **specific heat** of the substance. The specific heat represents the quantity of heat required to raise the temperature of one gram of the substance by one degree Celsius. (Specific heats for many substances are tabulated in handbooks of chemical data.) Although the specific heat is not constant over all temperatures, it remains constant for many substances over fairly broad ranges of temperatures (such as in this experiment). Specific heats are quoted in units of kilojoules per gram per degree, kJ/g-°C (or in molar terms, in units of kJ/mol-°C).

The calorimeter for this experiment is pictured in Figure 9-1. The calorimeter consists of two nested plastic foam coffee cups and a cover, with thermometer and stirring wire inserted through holes punched in the cover. As you know, plastic foam does not conduct heat well and will not allow heat generated by a chemical reaction to be lost to the room. (Coffee will not cool off as quickly in such a cup as in a china or paper cup.) Other sorts of calorimeters might be available in your laboratory; your instructor will demonstrate such calorimeters if necessary. The simple coffee-cup calorimeter generally gives acceptable results, however.

Experiment 9:   Calorimetry

Although the coffee cups used in this experiment do not conduct heat well, they do still transfer some heat. In addition, a small quantity of heat may be transferred to or from the metal wire used for stirring the calorimeter's contents or to the thermometer used to measure temperature changes. Rather than determining the influence of each of these separately, a function called the **calorimeter constant** can be determined for a given calorimeter in advanced practice. The calorimeter constant represents what portion of the heat flow from a chemical or physical process conducted in the calorimeter goes to the apparatus itself, rather than affecting the temperature of the water in the calorimeter. Once the calorimeter constant has been determined for a given apparatus, the value determined can be applied whenever that calorimeter is employed in subsequent experiments. Because plastic foam does not transfer much heat, and because most thermometers now contain colored hydrocarbons rather than mercury metal, we will ignore the quantity of heat transferred to or from the calorimeter itself in this experiment.

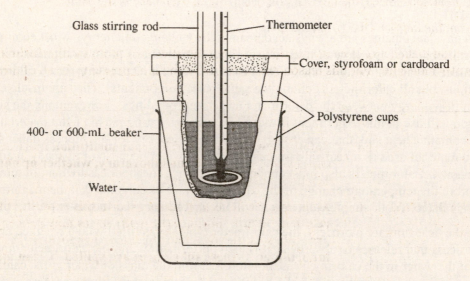

Figure 9-1. A simple calorimeter made from two nested plastic foam coffee cups
Make certain the stirring wire can be agitated easily. A glass beaker should be used to support the plastic cups to prevent their tipping over.

In this experiment you will measure the heat flows for two acid-base neutralization reactions. Many chemical reactions are performed routinely with the reactant species dissolved in water. The use of such solutions has many advantages over the use of "dry chemical" methods. The presence of a solvent matrix permits easier and more intimate mixing of the reactant species, solutions may be measured by volume rather than mass, and the presence of water may act as a moderating agent in the reaction.

Heats of reaction between species dissolved in water are especially easy to measure because the measurement may be performed in a simple calorimeter as described above. A measured volume of solution of one of the reagents is placed in the calorimeter cup and the temperature is determined. The second reagent solution is prepared in a separate container and is allowed to come to the same temperature as the solution in the calorimeter. When the two solutions have come to the same temperature, they are combined in the calorimeter, and the temperature of the calorimeter contents is monitored as the reaction takes place. If the chemical reaction is **exergonic**, the temperature of the water in the calorimeter will increase as energy is transferred to it from the reagents. If the chemical reaction is **endergonic**, the temperature of the water in the calorimeter will decrease as thermal energy is drawn from the water into the reactant substances.

For the purpose of tabulating heats of reaction in the chemical literature, such heats are usually converted to the basis of the number of kilojoules of heat energy that flows in the reaction per mole of reactant (or product). A typical experimental determination may only use a small fraction of a mole of reactant, and so only a few joules of heat energy would be involved in the experiment, but the results are converted to the basis of one mole. When such a determination of heat flow is conducted in a calorimeter that is equilibrated with the constant pressure of the atmosphere, the heat flow in kilojoules per mole is given the symbol ΔH, and is referred to as the **enthalpy change** for the reaction. Handbooks of chemical data list the enthalpy changes for many reactions.

You will measure the heat of reaction for two "different" reactions:

$$HCl + NaOH \rightarrow$$

$$HNO_3 + KOH \rightarrow$$

Both of these reactions represent the neutralization of an acid by a base, and although the reactions appear to involve different substances, the net reaction occurring in each case is the same:

$$H^+ \text{ (from the acid)} + OH^- \text{ (from the base)} \rightarrow H_2O$$

The actual reaction that occurs in each situation is the combination of a proton with a hydroxide ion, producing a water molecule. For this reason, the heat flows should be the same for all of these reactions.

Safety Precautions	• Safety eyewear approved by your institution must be worn at all times while you are in the laboratory, whether or not you are working on an experiment.
	• Although the acids and bases used in this experiment are in relatively dilute solution, the acids/bases may concentrate if they are spilled on the skin and the water evaporates from them. Wash immediately if these substances are spilled. Clean up all spills on the benchtop.

Apparatus/Reagents Required

- calorimeter/thermometer/stirrer apparatus (see Figure 9-1) supported in glass beaker
- 2.0 M hydrochloric acid
- 2.0 M nitric acid
- 2.0 M sodium hydroxide
- 2.0 M potassium hydroxide,

Procedure

Set up the calorimeter apparatus in a glass beaker that can accommodate and support the plastic cups.

Obtain 75 mL of 2.0 M NaOH and place it in the calorimeter. Record the exact concentration indicated on the reagent bottle. Obtain 75 mL of 2.0 M HCl in a clean, dry beaker and record the exact concentration indicated on the reagent bottle.

Allow the two solutions to stand until their temperatures are the same (within $\pm\ 0.5°C$). Be sure to rinse off and dry the thermometer when transferring between the solutions to prevent premature mixing of the reagents. Record the temperature(s) of the solutions to the nearest 0.2°C.

Add the HCl from the beaker *all at once* to the calorimeter, cover the calorimeter quickly, stir the mixture for 30 seconds, and record the highest temperature reached by the mixture (to the nearest 0.2°C).

From the change in temperature undergone by the mixture upon reaction and the total mass (volume) of the combined solutions, calculate the quantity of heat that flowed from the reactant species into the water of the solution.

Calculate the number of moles of water produced when 75 mL of 2.0 M HCl reacts with 75 mL of 2.0 M NaOH. Calculate ΔH in terms of the number of kilojoules of heat energy transferred when 1 mol of water is formed in the neutralization of aqueous HCl with aqueous NaOH.

Example:

Suppose 50.0 ml of 3.01 M HCl solution at 25.2°C is added to 50.0 mL of 3.00 M NaOH solution also at 25.2°C, whereupon it is observed that the temperature of the mixture rises to a maximum of 45.6°C. Calculate the heat flow, Q, for this experiment and ΔH for the process. Assume that the density of the solutions is 1.00 g/mL and that the specific heat capacity of the solution is 4.184 J/g°C

The temperature change undergone by the system is $\Delta T = (45.6°C - 25.2°C) = 20.4°C$

In general, $Q = mC\Delta T = (100.\ \text{g})(\dfrac{4.184\ \text{J}}{\text{g}\ °C})(20.4°C) = 8535\ (8.54 \times 10^3)\ \text{J} = 8.54\ \text{kJ}$

Since the reaction is of 1:1 stoichiometry, 50.0 mL of ~3 M HCl (or NaOH) reacting will result in the production of 0.150 mol of water

Therefore, $\Delta H = -\dfrac{8.54\ \text{kJ}}{0.150\ \text{mol}} = -56.9\ \text{kJ/mol}$ (heat is evolved).

Repeat the determination of ΔH for the HCl/NaOH reaction twice, and calculate a mean value for ΔH for the reaction from your three determinations.

Repeat the process above for the reaction between $HNO_3(aq)$ and $KOH(aq)$, making a total of three determinations of the heat flow for that reaction.

Compare the values of ΔH for the two neutralization reactions. Why is ΔH the same for both reactions?

EXPERIMENT 10

Precipitation Reactions

Objective

Reactions in which a solid is produced when aqueous solutions of ionic solutes are mixed are classified as **precipitation** reactions. Several examples of such reactions will be investigated in this experiment.

Introduction

There are more than 100 known elements and millions of known compounds. Throughout history, chemists have sought to organize the wealth of data and observations that have been recorded for these substances. One method of organizing such information is to classify reactions into groups according to some characteristic the reactions have in common. For example, reactions resulting in the formation of a precipitate when the reagents are mixed constitute one possible smaller group, and are the subject of this experiment.

It should be realized, however, that dividing all the possible chemical reactions into smaller groups for study is a somewhat arbitrary method: it is possible, for example, for the same reaction to belong to two or more groups simultaneously. Such classification schemes are intended to be a means of simplifying the study of chemical reactions. Such classification schemes are not necessarily absolute.

Certain substances are not very soluble in water. Such substances are frequently generated *in situ* in a reaction vessel by the addition of various other substances that are themselves very soluble. For example, silver chloride is not soluble in water. If an aqueous solution of silver nitrate (very soluble) is mixed with an aqueous solution of sodium chloride (very soluble), the combination of silver ions from one solution and chloride ions from the other solution generates silver chloride, which then forms a precipitate that settles to the bottom of the container. The solution that remains above the precipitate of silver chloride effectively becomes a solution of sodium nitrate. Silver ions and sodium ions have switched partners, ending up in a compound with the negative ion that originally came from the other substance:

$$AgNO_3(aq) + NaCl(aq) \rightarrow AgCl(s) + NaNO_3(aq)$$

The silver ion and sodium ion have replaced each other in this process; this sort of reaction is sometimes called a **double displacement reaction**.

To clarify what is really happening in such reactions, it is often more instructive to write the reaction equation in its **net ionic** form. In a net ionic equation for a precipitation reaction, only the ions involved in actually forming the precipitate are shown; the other ions contained in the original reagents used are called **spectator ions**, and are still present in solution after the precipitate has formed. For example, for the reaction of silver nitrate and sodium chloride given earlier, the net ionic reaction is:

$$Ag^+(aq) + Cl^-(aq) \rightarrow AgCl(s)$$

The net ionic reaction is especially instructive because it implies that any solution containing silver ion should react with any solution containing chloride ion, since it is these ionic species that are really reacting. For example, if dilute hydrochloric acid, HCl, were added to a solution of silver nitrate, a precipitate would be expected to form. The reaction of silver nitrate with sodium chloride, and that of silver nitrate with hydrochloric acid, are exactly the same when the net ionic reaction is considered.

Your textbook also lists some general rules for the solubility of ionic compounds, which summarize which types of compounds are likely to be soluble and which are not. Those rules are paraphrased here:

1. Most nitrate salts are readily soluble in water.

2. Most sodium, potassium and ammonium salts are readily soluble in water.

3. Most chloride salts are soluble in water, with three common exceptions: $AgCl$, $PbCl_2$ and Hg_2Cl_2.

4. Most sulfate salts are soluble in water, with three common exceptions: $BaSO_4$, $PbSO_4$ and $CaSO_4$.

5. Most hydroxide compounds are not soluble or are only slightly soluble. Soluble exceptions are $NaOH$ and KOH. Slightly soluble exceptions are $Ca(OH)_2$ and $Ba(OH)_2$.

6. Most salts containing sulfide, carbonate or phosphate ions are insoluble or only slightly soluble.

Safety Precautions	• **Safety eyewear approved by your institution must be worn at all times while you are in the laboratory, whether or not you are working on an experiment.**
	• **Salts of the heavy metals are toxic if ingested. Some are corrosive or irritating to the skin. Wash immediately after using these reagents.**
	• **Some of the reagents used in this experiment may stain skin and clothing. Wear plastic gloves during this experiment.**
	• **Hydrochloric and sulfuric acids are damaging to skin, eyes and clothing. Wash immediately if spilled. Wash after using these substances. Inform the instructor of any spills.**
	• **Some of the materials used in this experiment are environmental hazards. Do not pour down the drain. Dispose of the materials as directed by the instructor.**

Apparatus/Reagents Required

- 96-well plastic microtiter plate

- 5 or 6 plastic micropipets

- spot plate or semi-micro test tubes

- 0.1 M solutions of the following solutes: sulfuric acid, hydrochloric acid, silver nitrate, sodium carbonate, sodium chloride, potassium chromate, lead(II) acetate, sodium sulfate, barium chloride, ammonium chloride, copper(II) sulfate

- plastic gloves

Procedure

Record all data and observations directly on the report pages in ink.

Since some of the reagents in this experiment may stain skin, you should wear gloves during this experiment.

Obtain 1 mL samples of the reagents listed, using a spot plate or semi-micro test tubes to store the solutions: sulfuric acid, hydrochloric acid, silver nitrate, sodium carbonate, sodium chloride, potassium chromate, lead(II) acetate, sodium sulfate, barium chloride.

In the following reactions, use a clean 96-well microtiter plate as a reaction chamber and several micropipets to dispense the reagents. Combine the reagent solutions as indicated in the list that follows.

Rinse the micropipets with distilled water before transferring them between different solutions. To rinse the pipet without generating a lot of waste material, do the following. First, *empty* the pipet of the previous reagent. Then, draw water into the pipet and rotate the pipet to rinse the inside walls with the water. Discard this rinse water into a waste beaker and repeat the rinsing.

Record whether or not a precipitate forms for each combination of solutes. Also record the color of the solutions used and of the precipitates that form.

Use the solubility rules from your textbook to determine what precipitate forms in each reaction.

Write balanced overall and net ionic equations for each reaction. If no precipitate forms, indicate, based on the solubility rules, why no precipitate is likely for the particular reagents involved.

Reactions:

silver nitrate and sodium chloride $\rightarrow$

silver nitrate and hydrochloric acid $\rightarrow$

silver nitrate and sulfuric acid $\rightarrow$

silver nitrate and sodium sulfate $\rightarrow$

silver nitrate and sodium carbonate $\rightarrow$

silver nitrate and potassium chromate $\rightarrow$

silver nitrate and lead(II) acetate $\rightarrow$

silver nitrate and barium chloride $\rightarrow$

potassium chromate and lead(II) acetate $\rightarrow$

potassium chromate and barium chloride $\rightarrow$

potassium chromate and copper(II) sulfate $\rightarrow$

lead(II) acetate and sodium carbonate $\rightarrow$

lead(II) acetate and sodium chloride $\rightarrow$

lead(II) acetate and sodium sulfate $\rightarrow$

lead(II) acetate and sulfuric acid $\rightarrow$

lead(II) acetate and barium chloride $\rightarrow$

lead(II) acetate and hydrochloric acid $\rightarrow$

Experiment 10:   Precipitation Reactions

lead(II) acetate and copper(II) sulfate →

barium chloride and sodium carbonate →

barium chloride and sodium sulfate →

barium chloride and sulfuric acid →

barium chloride and copper(II) sulfate →

sodium chloride and sodium carbonate →

ammonium chloride and lead(II) acetate →

ammonium chloride and silver nitrate →

ammonium chloride and copper(II) sulfate →

Name: _____ Section: _____

Lab Instructor: _____ Date: _____

EXPERIMENT 10

Precipitation Reactions

Pre-Laboratory Questions

1. Use your textbook or an encyclopedia of chemistry to define each of the following terms:

    Precipitation reaction

    _____

    _____

    _____

    _____

    _____

    _____

    _____

    Spectator ions

    _____

    _____

    _____

    _____

    _____

    _____

    Net ionic equation for a reaction

    _____

    _____

    _____

    _____

    _____

    _____

2. Chapter 7 lists general solubility rules for ionic solutes. The following exercises are *mis*statements of those solubility rules. Explain what the error is in each statement and correct it.

    a. All sodium, potassium and ammonium salts are not appreciably soluble in water.

        _____

        _____

        _____

        _____

        _____

    b. Most carbonate, sulfide and phosphate salts readily dissolve in water.

        _____

        _____

        _____

        _____

        _____

    c. Most hydroxide compounds are readily soluble in water, with the exception of NaOH and KOH which are not appreciably soluble in water.

        _____

        _____

        _____

        _____

        _____

3. Based on the solubility rules given in Chapter 7 in your textbook, write the name and formula of 10 substances that you predict would *not* be appreciably soluble in water.

    _____

    _____

    _____

    _____

    _____

    _____

    _____

    _____

    _____

    _____

    _____

EXPERIMENT 10

Precipitation Reactions

Results/Observations

Observations on mixing: include the color and formula of any precipitates formed.

Silver nitrate and sodium chloride _____

Silver nitrate and hydrochloric acid _____

Silver nitrate and sulfuric acid_____

Silver nitrate and sodium sulfate _____

Silver nitrate and sodium carbonate_____

Silver nitrate and potassium chromate _____

Silver nitrate and lead(II) acetate _____

Silver nitrate and barium chloride_____

Potassium chromate and lead(II) acetate _____

Potassium chromate and barium chloride _____

Potassium chromate and copper(II) sulfate _____

Lead(II) acetate and sodium carbonate _____

Lead(II) acetate and sodium chloride _____

Lead(II) acetate and sodium sulfate _____

Lead(II) acetate and sulfuric acid _____

Lead(II) acetate and barium chloride _____

Lead(II) acetate and hydrochloric acid _____

Lead(II) acetate and copper(II) sulfate _____

Barium chloride and sodium carbonate _____

Barium chloride and sodium sulfate _____

Barium chloride and sulfuric acid _____

Barium chloride and copper(II) sulfate _____

Sodium chloride and sodium carbonate _____

Ammonium chloride and lead(II) acetate _____

Ammonium chloride and silver nitrate _____

Ammonium chloride and copper(II) sulfate _____

Net ionic equations: Write the net ionic equation for each of the reactions you performed that resulted in the formation of a precipitate.

_____

_____

_____

_____

_____

_____

_____

_____

_____

_____

_____

_____

Questions

1.   This experiment discusses general, qualitative rules about the solubility of ionic substances. The solubility of a substance is determined on a quantitative basis through a function called the solubility product for the substance. Find a definition in your textbook or online for the solubility product.

_____

_____

_____

_____

_____

2.   Complete and balance the following reactions in which a precipitate forms:

  a. $K_2SO_4(aq) + Ba(NO_3)_2(aq) \rightarrow$ _____

  b. $H_2S(aq) + CuSO_4(aq) \rightarrow$ _____

  c. $FeCl_3(aq) + KOH(aq) \rightarrow$ _____

  d. $Na_2CO_3(aq) + CuSO_4(aq) \rightarrow$ _____

  e. $CuCl_2(aq) + (NH_4)_2S(aq) \rightarrow$ _____

  f. $Na_3PO_4(aq) + AlCl_3(aq) \rightarrow$ _____

Properties and Reactions of Acids and Bases

Objective

Reactions in aqueous solution in which protons are transferred between species are called **acid-base** reactions. Several aspects of acid-base chemistry will be qualitatively studied in this experiment.

Introduction

There are many theories and definitions that attempt to explain what constitutes an acid or a base. An early and still useful theory, developed by Arrhenius in the late 1800s, defined an acid as a substance that produces hydrogen ions, H^+, when dissolved in water. A base, according to Arrhenius, is a species that produces hydroxide ions, OH^-, when dissolved in water. For example, hydrogen chloride is an acid in the Arrhenius theory because hydrogen chloride ionizes when dissolved in water, producing hydrogen ions.

$$HCl \rightarrow H^+ + Cl^-$$

In particular, hydrogen chloride is called a *strong acid* because virtually every HCl molecule ionizes when dissolved. Other substances, although they do in fact produce hydrogen ions when dissolved, do not completely dissociate when dissolved, and are called *weak acids*. For example, the acidic component of vinegar (acetic acid) is a weak acid

$$CH_3COOH \rightleftharpoons H^+ + CH_3COO^-$$

The double-ended arrow in the equation for the ionization of acetic acid indicates that this substance reaches equilibrium when dissolved in water, at which point a certain fixed concentration of hydrogen ion is present. The concentration of hydrogen ion produced by dissolving a given amount of weak acid is typically several orders of magnitude less than if the same amount of strong acid is dissolved.

In a similar manner, there are both strong and weak bases in the Arrhenius scheme. Sodium hydroxide, for example, is a strong base:

$$NaOH(s) \rightarrow Na^+(aq) + OH^-(aq)$$

For every mole of NaOH that might be dissolved in water, an equivalent number of moles of OH^- is produced. In contrast, ammonia, NH_3, is a weak base:

$$NH_3(g) + H_2O \rightleftharpoons NH_4^+(aq) + OH^-(aq)$$

Ammonia, in effect, reacts with water and comes to equilibrium, at which point a certain fixed concentration of OH^- exists. As with weak acids, the concentration of hydroxide ion in a solution of a weak base is much smaller than if the same amount of strong base had been dissolved.

The most important reaction of acids and bases is *neutralization*. The hydrogen ion from an aqueous acid will combine with the hydroxide ion from an aqueous base, producing water and a salt. For example,

$$HCl + NaOH \rightarrow NaCl + H_2O$$

$$HNO_3 + NaOH \rightarrow NaNO_3 + H_2O$$

The net reaction in each of these equations is the same, and it is typical of the reaction between acids and bases in aqueous solution

$$H^+ + OH^- \rightarrow H_2O$$

Although the Arrhenius definitions of acids and bases have proved very useful, the theory is restricted to the situation of aqueous solutions. Aqueous solutions are most common, but it is reasonable to question if HCl and NaOH will still behave as an acid or base when dissolved in some other solvent. The Brønsted-Lowry theory of acids and bases extends the Arrhenius definitions to more general situations. An acid is defined in the Brønsted-Lowry theory as any species that provides hydrogen ions, H^+, in a reaction. Since a hydrogen ion is nothing more than a simple proton, acids are often referred to as being "proton donors" in the Brønsted-Lowry scheme. Bases in the Brønsted-Lowry system are defined as any species that receives hydrogen ions from an acid. Bases are referred to then as "proton acceptors." The similarity between the definitions of an acid in the Arrhenius and Brønsted-Lowry schemes is obvious. On first glance, however, it would seem that the definition of what constitutes a base differs between the two systems. Realize that hydroxide ion in aqueous solution (Arrhenius definition) will very readily accept protons from any acid that might be added. Therefore, hydroxide ion is in fact a base in the Brønsted-Lowry system as well.

Safety Precautions	• Safety eyewear approved by your institution must be worn at all times while you are in the laboratory, whether or not you are working on an experiment.
	• The acids and bases used in this experiment may be damaging to eyes, skin and clothing. The acids are especially dangerous if the solutions are allowed to concentrate by evaporation. Wash immediately if the acids or bases are spilled. Wash after using. Inform the instructor of any spills.

Apparatus/Reagents Required

- 0.1 *M* solutions of the following acids and bases: hydrochloric acid, acetic acid, sodium hydroxide, ammonia
- 6 *M* solutions of HCl and NaOH
- conductivity tester
- universal indicator and color chart
- toothpicks
- pH paper (extended range: pH 0–14)
- watch glass
- semimicro test tubes

Procedure

1. pH of Acid-Base Solutions

Obtain 10 drops each of 0.1 M HCl and 0.1 M acetic acid in separate clean semi-micro test tubes. Using extended-range pH paper, determine the pH of each solution.

How do you account for the fact that the pH of the acetic acid solution is higher than that of the HCl solution, although both are at the same molar concentration (0.1 M)?

Obtain 10 drops each of 0.1 M NaOH and 0.1 M ammonia in separate clean semi-micro test tubes. Using extended-range pH paper, determine the pH of each solution.

How do you account for the fact that the pH of the ammonia solution is lower than that of the NaOH solution, although both are at the same molar concentration (0.1 M)?

2. Conductivity of Acid-Base Solutions

One easy method for demonstrating the difference in the degree of ionization between strong acid and weak acid solutions (or of other electrolytes) is to measure the relative electrical conductivity of the solutions. Solutions of strong acids (or other strongly ionized electrolytes) conduct electrical currents well, whereas solutions of weak acids (or other weak electrolytes) conduct electricity poorly.

Your instructor has set up a light-bulb conductivity tester and will demonstrate the electrical conductivity of the 0.1 M acid and base solutions. The light-bulb will glow brightly when the electrodes of the conductivity tester are immersed in strong acid or strong base solution, but will glow only dimly when immersed in solutions of weak acid or weak base.

Record which of the acids/bases are strong electrolytes, and which are weak.

3. Production of Salts from Acid-Base Reactions

Combine five drops of 6 M HCl and five drops of 6 M NaOH on a clean, dry watch glass. Stir the mixture with a toothpick.

Allow the mixture to evaporate over the remainder of the laboratory period. Before leaving the laboratory, examine the watch glass. What is the identity of the solid residue remaining on the watch glass?

4. Neutralization

Obtain 2.0±0.1 mL of 0.1 M HCl in a clean test tube, and add one drop of universal indicator. Record the color of the indicator and the pH of the solution. This indicator changes color gradually with pH and can be used to monitor the pH of a solution during a neutralization reaction. Keep handy the color chart provided with the indicator. It shows the color of the indicator under different pH conditions.

Obtain a sample of 0.1 M NaOH in a small beaker, and begin adding NaOH to the sample of HCl and indicator one drop at a time with a medicine dropper or plastic micropipet. Record the color and pH after three drops of NaOH have been added.

Continue adding NaOH dropwise, recording the color and pH at three-drop increments until the pH has risen to pH greater than 10. Approximately how many drops of NaOH were required to reach pH 7 (neutral)?

Given that 1 mL of liquid is approximately equivalent to 20 average drops, approximately how many mL of 0.1 M NaOH were required to neutralize the 2-mL sample of 0.1 M HCl?

Repeat the process, using 2 mL of 0.1 M acetic acid in place of the HCl. How does the change in pH during the addition of NaOH differ when a weak acid is neutralized?

5. Heat of Neutralization

Obtain approximately 5 mL each of 6 M HCl and 6 M NaOH in separate clean test tubes. Determine the temperature of each solution, *being sure to wipe the thermometer with a paper towel before switching between solutions.*

Support the tube containing the HCl in a test tube rack and insert the thermometer. Quickly pour the NaOH solution into the HCl solution, stir briefly, and monitor the thermometer reading. Record the highest temperature reached by the mixture. Is the neutralization reaction exothermic or endothermic?

Repeat the process using 5 mL 6 M acetic acid in place of the HCl.

Name: _____ Section: _____

Lab Instructor: _____ Date: _____

EXPERIMENT 11

Properties and Reactions of Acids and Bases

Pre-Laboratory Questions

1. Two major models for acid-base behavior are the Arrhenius model and the Brønsted-Lowry model. How are these two models similar? How do they differ? Give an example of a situation in which a substance would be classified as an acid under the Brønsted-Lowry model, but not under the Arrhenius model.

    _____

    _____

    _____

    _____

    _____

    _____

    _____

    _____

    _____

2. Write the formula for the Brønsted-Lowry base for each of the following acids.

    HNO_3 _____

    H_2O _____

    HSO_4^- _____

    NH_3 _____

    H_2SO_4 _____

3. Write the formula for the Brønsted-Lowry acid for each of the following bases.

    NH_2^- _____

    HSO_4^- _____

    OH^- _____

    NH_3 _____

    H_2O _____

4. Define neutralization and give three examples of balanced chemical equations for neutralization reactions.

_____

_____

_____

_____

_____

_____

_____

_____

_____

_____

5. Suppose the salts listed in the table below were to be prepared by reaction of acids with bases. Tell what acid and what base would be most appropriate for preparing each salt.

salt	acid	base
sodium chloride	_____	_____
sodium nitrate	_____	_____
potassium sulfate	_____	_____
sodium perchlorate	_____	_____
lithium acetate	_____	_____

EXPERIMENT 11

Properties and Reactions of Acids and Bases

Results/Observations

1. pH of Acid-Base Solutions

Solution	pH measured	Solution	pH measured
0.1 M HCl	_____	0.1 M CH$_3$COOH	_____
0.1 M NaOH	_____	0.1 M NH$_3$	_____

   Explanation of each observed pH

   _____

   _____

   _____

   _____

2. Conductivity of Acid-Base Solutions

Solution	Observed conductivity
0.1 M HCl	_____
0.1 M CH$_3$COOH	_____
0.1 M NaOH	_____
0.1 M NH$_3$	_____

3. Production of Salts from Acid-Base Reactions

   Observation of HCl/NaOH product

   _____

   _____

   _____

4. Neutralization

   HCl

   Initial color of universal indicator_____ pH of solution _____

   Color of indicator after three drops NaOH _____ pH of solution _____

   Drops of NaOH required to reach pH 7 _____ Drops to reach pH 10 _____

   mL NaOH required to reach pH 7 _____

CH₃COOH

Initial color of universal indicator_____   pH of solution _____

Color of indicator after three drops NaOH _____   pH of solution _____

Drops of NaOH required to reach pH 7 _____   Drops to reach pH 10 _____

mL NaOH required to reach pH 7 _____

5.   Heat of Neutralization

HCl

Initial temperature of acid _____°C

Highest temperature reached on adding NaOH _____°C

Is reaction exothermic or endothermic? _____

CH₃COOH

Initial temperature of acid _____°C

Highest temperature reached on adding NaOH _____°C

Is reaction exothermic or endothermic?_____

Questions

1.   Write balanced chemical equations for the reaction between HCl and NaOH and between CH₃COOH with NaOH.

_____

_____

2.   Would you expect the pH of 0.25 *M* acetic acid to be higher or lower than the pH of 0.25 *M* hydro-chloric acid solution? Explain.

_____

_____

_____

_____

3.   In Part 3 of the procedure you mixed equal amounts of HCl and NaOH, and allowed the water of the solutions to evaporate. What is the identity of the solid residue that remained? Suppose H₂SO₄ had been used instead of HCl: what residue would remain on evaporation in this case?

_____

_____

_____

_____

Oxidation-Reduction Reactions

Objective

Oxidation-reduction reactions involve a transfer of electrons from one species to another, and form one of the main classes of chemical reactions. Several examples of such processes will be examined in this experiment.

Introduction

A large and important group of common chemical reactions can be classified as **oxidation-reduction** (or **redox**) processes. Oxidation-reduction processes involve the transfer of electrons from one species to another. For example, in the reaction

$$Zn(s) + CuSO_4(aq) \rightarrow ZnSO_4(aq) + Cu(s)$$

$Zn(s)$ represents uncharged elemental zinc atoms, whereas in $ZnSO_4(aq)$, the zinc exists in solution as $Zn^{2+}(aq)$ ions. Each zinc atom has effectively lost two electrons in the process and is said to have been oxidized:

$$Zn(s) \rightarrow Zn^{2+}(aq) + 2e^- \qquad \textit{oxidation}$$

Oxidation is defined as a loss of electrons by an atom or ion, or better, as a transfer of electrons from one species to another. Similarly, in the preceding reaction, $Cu^{2+}(aq)$ ions in solution have each gained two electrons, becoming uncharged elemental copper atoms:

$$Cu^{2+}(aq) + 2e^- \rightarrow Cu(s) \qquad \textit{reduction}$$

Copper ions are reduced in the process. Reduction is defined as a gain of electrons by a species, or better, as the transfer of electrons to a species from another species.

To clarify the transfer of electrons between the species undergoing oxidation and the species undergoing reduction, redox reactions are usually divided into two half-reactions, one for each process – as was demonstrated in the zinc/copper reaction above. Some redox reactions and their corresponding half-reactions follow:

$$2HgO \rightarrow 2Hg + O_2 \qquad \textit{overall}$$
$$2Hg^{2+} + 4e^- \rightarrow 2Hg \qquad \textit{reduction}$$
$$2O^{2-} \rightarrow O_2 + 4e^- \qquad \textit{oxidation}$$

$$Cl_2 + 2I^- \rightarrow 2Cl^- + I_2 \qquad \textit{overall}$$
$$Cl_2 + 2e^- \rightarrow 2Cl^- \qquad \textit{reduction}$$
$$2I^- \rightarrow I_2 + 2e^- \qquad \textit{oxidation}$$

Oxidation-reduction reactions are very common and take many forms. Reactions in which elemental substances combine to form a compound, or in which compounds are decomposed to elemental substances (or simpler compounds), are redox reactions. The reactions of metallic substances with acids, or with corrosive gases, and the reaction of many species in solution are all examples of oxidation-reduction processes.

Oxidation-reduction is especially important in the study of electrochemical processes. For example, chemical batteries use a spontaneous oxidation-reduction process as a source of electrical energy. Oxidation-reduction also is used commonly to electroplate one metal on the surface of another, in jewelry-making, for example. The study of redox reactions is also important in the study of the corrosion of metals (oxidation of the metal by O_2 in the atmosphere) and in its prevention. Finally, the study of oxidation-reduction reactions is of the utmost importance in biochemistry, since many of the chemical reactions in a living cell – especially in metabolism – are redox processes.

Safety Precautions	• **Safety eyewear approved by your institution must be worn at all times while you are in the laboratory, whether or not you are working on an experiment.** • **The flame produced by burning magnesium is intensely bright and can damage the eyes. Do not look directly at the flame of burning magnesium.** • **Hydrogen peroxide solutions are unstable and will burn the skin if spilled.** • **Lead compounds are extremely toxic if ingested. Wash hands and arms after using them.** • **Potassium permanganate will stain skin and clothing if spilled.** • **"Chlorine water" may burn the skin and may emit toxic chlorine gas. Wash after use, and confine the use to the fume exhaust hood.** • **Methylene chloride is toxic if inhaled or absorbed through the skin. Wash after use. Confine its use to the fume exhaust hood.** • **Some of the substances used in this experiment are environmental hazards. Do not pour down the drain. Dispose of these substances as directed by the instructor.**

Apparatus/Reagents Required

- oxygen gas (*instructor demonstration*)
- magnesium ribbon
- steel wool
- 3% hydrogen peroxide
- 0.5 *M* copper(II) sulfate
- 1 *M* lead acetate
- metallic zinc strips

- metallic copper strips
- 10% potassium iodide
- 10% potassium bromide
- chlorine water
- methylene chloride

Procedure

1. Oxidation of Magnesium

Obtain a one-inch strip of magnesium ribbon. When magnesium is ignited, the flame produced is dangerous to the eyes. *Caution*! Do not look directly at burning magnesium.

Hold the magnesium ribbon with tongs, and have ready a beaker of distilled water to catch the product of the reaction as it is formed.

Ignite the magnesium ribbon in a burner flame, and hold the burning ribbon over the beaker of water. The unbalanced reaction is

$$Mg + O_2 \rightarrow MgO$$

Allow the beaker of water containing the magnesium oxide produced by the reaction to stand for 5–10 minutes; then test the water with pH paper. Why is the solution basic? Magnesium is oxidized in the reaction ($0 \rightarrow +2$), whereas oxygen is reduced ($0 \rightarrow -2$).

2. Oxidation of Iron

Your instructor will demonstrate (in the exhaust hood) the burning of iron (steel wool) in a stream of pure oxygen. Iron oxidizes only slowly in air at room temperature (forming rust), but the reaction is considerably faster when using pure oxygen and elevated temperatures. Iron has two common oxidation states (+2, +3). The unbalanced reactions are

$$Fe + O_2 \rightarrow FeO \quad and \quad Fe + O_2 \rightarrow Fe_2O_3$$

3. Oxidation-Reduction of Hydrogen Peroxide

Place about 20 mL of 3% hydrogen peroxide in a large test tube. Obtain a single crystal of potassium permanganate.

Ignite a wooden splint, then blow out the flame so that the splint is still glowing, and then quickly add the crystal of potassium permanganate to the hydrogen peroxide sample to initiate the decomposition of the hydrogen peroxide.

Using forceps, insert the glowing wood splint halfway into the test tube. The glowing wood splint should immediately burst into flame as it comes in contact with the pure oxygen being generated by the decomposition reaction (unbalanced):

$$H_2O_2 \rightarrow H_2O + O_2$$

4. Oxidation-Reduction of Copper and Zinc

Obtain about 5 mL of 0.5 *M* copper(II) sulfate solution in a small test tube. Add a thin strip of metallic zinc and allow the solution to stand for 15 to 20 minutes. Then remove the zinc strip and examine the coating that has formed on the strip. The reaction is

$$Zn(s) + CuSO_4(aq) \rightarrow ZnSO_4(aq) + Cu(s)$$

5. Oxidation-Reduction of Lead and Zinc

Obtain about 10 mL of 1 *M* lead acetate solution in a test tube, and place the test tube in a place where it will not be disturbed.

Add a thin strip of metallic zinc to the solution. Allow the solution to stand for 30 to 40 minutes.

After this period examine the "tree" of metallic lead that has formed. The unbalanced reaction is

$$Zn(s) + Pb(CH_3COO)_2(aq) \rightarrow Pb(s) + Zn(CH_3COO)_2(aq)$$

6. Oxidation-Reduction of Copper and Silver

Scrub a strip of metallic copper with steel wool and rinse with water. Place 10 to 15 mL of 0.1 *M* silver nitrate solution in a test tube, and add the copper strip.

Allow the test tube to stand for 15 to 20 minutes, then examine the copper strip. The unbalanced reaction equation is

$$Cu(s) + AgNO_3(aq) \rightarrow Cu(NO_3)_2(aq) + Ag(s)$$

7. Oxidation-Reduction of the Halogens

Obtain about 1 mL each of 10% potassium iodide and 10% potassium bromide in separate test tubes.

Add 1 mL of "chlorine water" to each test tube; stopper and shake briefly. Record any color changes.

In the fume exhaust hood, add approximately 10 drops of methylene chloride to each test tube, stopper, and shake.

Elemental iodine and elemental bromine have been produced by replacement; these substances are more soluble in methylene chloride than in water and are preferentially extracted into this solvent. The colors of the methylene chloride layer are characteristic of these halogens. The unbalanced reactions are

$$Cl_2 + KI(aq) \rightarrow KCl(aq) + I_2$$

$$Cl_2 + KBr(aq) \rightarrow KCl(aq) + Br_2$$

Name: _____ Section: _____

Lab Instructor: _____ Date: _____

EXPERIMENT 12

Oxidation-Reduction Reactions

Pre-Laboratory Questions

1. Oxidation may be defined as a loss of electrons by an atom, ion or molecule, or as an increase in the oxidation state of an element in such a species. Give three examples that illustrate the equivalence of these two definitions.

    _____

    _____

    _____

    _____

    _____

    _____

    _____

2. Identify the element being oxidized and the element being reduced in each of the following reactions

    a. $2K(s) + S(s) \rightarrow K_2S(s)$

        oxidized _____ reduced _____

    b. $Cu(s) + Cl_2(g) \rightarrow CuCl_2(s)$

        oxidized _____ reduced _____

    c. $2S(s) + 3O_2(g) \rightarrow 2SO_3(s)$

        oxidized _____ reduced _____

    d. $FeS(s) + H_2(g) \rightarrow Fe(s) + H_2S(g)$

        oxidized _____ reduced _____

    e. $2Cr_2S_3(s) + 3O_2(g) \rightarrow 2Cr_2O_3(s) + 6S(s)$

        oxidized _____ reduced _____

3. Write and balance the chemical equation for the oxidation of I^- ion in aqueous solution by chlorine gas (Cl_2), producing elemental iodine (I_2) and chloride ion in aqueous solution.

    _____

    _____

    _____

4.   Potassium permanganate ($KMnO_4$) is a very powerful oxidizing agent. It is often used in analyses for iron(II) salts because it quantitatively and rapidly oxidizes Fe^{2+} to Fe^{3+}. Write and balance the oxidation-reduction equation for the reaction of the permanganate ion (MnO_4^-) with Fe^{2+} in acidic aqueous solution, producing Mn^{2+} ion and Fe^{3+} ion.

_____

_____

_____

_____

_____

_____

_____

Name: _____ Section: _____

Lab Instructor: _____ Date: _____

EXPERIMENT 12

Oxidation-Reduction Reactions

Results/Observations

1. **Oxidation of Magnesium**

   Observations

   _____

   _____

   pH of solution _____ Explanation _____

   _____

2. **Oxidation of Iron**

   Observations

   _____

   _____

3. **Oxidation-Reduction of Hydrogen Peroxide**

   Observations

   _____

   _____

4. **Oxidation-Reduction of Copper and Zinc**

   Observations

   _____

   _____

5. **Oxidation-Reduction of Lead and Zinc**

   Observations

   _____

   _____

6. **Oxidation-Reduction of Copper and Silver**

   Observations

   _____

   _____

7.   **Oxidation-Reduction of the Halogens**

Observations

_____

_____

_____

_____

_____

Questions

1.   For each of the reactions performed, balance the given chemical equations. Indicate which atoms are oxidized and which are reduced in each of your equations.

_____

_____

_____

_____

_____

_____

_____

_____

2.   Oxidation-reduction reactions are very important in everyday life, as well as in the chemistry laboratory. Give three examples of oxidation-reduction reactions that might be made use of in everyday life: describe the reaction, what the reaction is used for and why it is important, and the chemical reactions involved (if known).

_____

_____

_____

_____

_____

_____

_____

_____

_____

EXPERIMENT 13

Counting by Weighing

Objective

Chemists determine how many individual atoms or molecules are in a sample of matter from the mass of the sample. This activity will give you some insight into how this calculation can be done.

Introduction

The concept of "counting by weighing" introduced in Chapter 8 of your textbook may be something new to you. Such an approach is a standard business practice, however, particularly when a business needs to take an inventory of small materials on hand. For example, if a company that makes carpentry nails had to count each individual item in their inventory the situation would be untenable. Rather than having to count each individual nail separately, the mass of a single nail is determined, and then the number of nails present in a bulk sample can be calculated easily. Similarly, in a high-volume "super" drugstore, a prescription for a large number of tablets or capsules of medication may be dispensed by this method.

In this experiment, you will demonstrate the process of "counting by weighing" for yourself using pennies as the item to be counted. This should give you some insight into why we can use average atomic masses in chemical reaction calculations. You will also investigate the effect of several isotopes on the average atomic mass of an element.

Safety Precautions	• **Safety eyewear approved by your institution must be worn at all times while you are in the laboratory, whether or not you are working on an experiment**

Apparatus/Reagents Required

- 25 pennies
- balance
- felt-tip pen

Procedure

1. Counting by Weighing

Obtain 25 pennies. In 1982, the composition of the United States penny coin was changed from one of nearly pure copper to a mostly zinc "sandwich" (the center of the coin is zinc, but the surfaces are layered in copper).

Separate the pennies into two piles: those that have dates through 1981, and those that have dates 1983 and following. Since pennies minted in 1982 may be of either type, set aside any such pennies and do not use them in the rest of the experiment.

With a felt-tip pen, number each of the remaining pennies so that you can identify them.

a. Pre-1982 Pennies

Weigh each of the pre-1982 pennies (to the nearest 0.01 g) and record the masses on the data page.

Calculate the average mass of your pre-1982 pennies.

To get a truly representative sample, it would be useful if all the students in the laboratory combined their data on the average mass of the pre-1982 pennies. On the chalkboard of the laboratory, write your name, the average mass you determined for the pre-1982 pennies, and the number of pre-1982 pennies you used.

When all the students in the class have recorded their average masses on the chalkboard, your instructor will demonstrate how to calculate the average mass of the pre-1982 penny using everyone's data. Record this average mass on the data page.

Based on the average mass of the pre-1982 penny as determined from all the students' data, what would be the mass of 55 pennies? How many pre-1982 pennies are contained in a pile of pennies that has a total mass of 310. g?

b. Post-1982 Pennies

You separated the pennies into two groups above, based on their date of minting, and then determined the average mass of the pre-1982 pennies only.

Although there may have been minor variations in the masses of the pre-1982 pennies due to different degrees of wear and tear, you should have found that most of the pennies had virtually the same mass. Post-1982 pennies, however, have masses that are considerably less.

Weigh each of your post-1982 pennies and record the masses on the data page.

Calculate the average mass of your post-1982 pennies.

On the chalkboard, write your name, the average mass of your post-1982 pennies, and the number of post-1982 pennies you used. After all the students in the lab have contributed their data on the post-1982 pennies, calculate the overall average mass of a post-1982 penny using all students' data.

Based on the average mass of the post-1982 penny, what would be the mass of 75 such pennies? How many such pennies would be contained in a pile of post-1982 pennies having a total mass of 250. g?

2. Effect of Isotopes on Average Atomic Masses

You have learned that most elements have several isotopic forms. The various isotopes of an element all have the same number of protons and electrons (so they are chemically the same), but differ in the number of neutrons present in the nucleus (which may result in slightly different physical properties).

You have also discussed how the average atomic mass listed for a particular element on the periodic table represents a weighted average of the masses of all the isotopes of the element. By "weighted average," we mean that the abundance of the element is reflected in the average atomic mass.

You can see what we mean by "weighted average" using the data you have collected for the pennies in Parts 1a and 1b above.

Using the individual masses of all the pennies (both pre- and post-1982), calculate the average mass of a penny (without regard to its minting date). Record your result.

You can arrive at the same average mass by another method of calculation, using the average mass you calculated for each type of penny, rather than the individual masses of all the pennies. Consider the following example:

A student has 5 pennies of average mass 3.11 g and 19 pennies of average mass 2.49 g. The weighted average mass of these 24 pennies is given by

$$\frac{\left[5(3.11\ g) + 19(2.49\ g)\right]}{24} = \frac{\left[15.55\ g + 47.31\ g\right]}{24} = 2.62\ g$$

The weighted average mass (2.62 g) is closer to 2.49 g than it is to 3.11 g because there were more pennies of the lower mass present in the sample: the weighted average has included the relative abundance of the two types of pennies.

Using the method outlined in the example above, calculate the weighted average mass of a penny. Record this average mass both on the data page and on the chalkboard.

How do the average masses reported by the students in your class compare? Are there significant differences in the average masses reported, or are they all comparable?

The concept of "weighted average" is also used by your school in calculating your grade point average. For example, if you receive an A in a 5-credit course, that counts more to your overall grade point average than would an A in a 1-credit course.

EXPERIMENT 13

Counting by Weighing

Pre-Laboratory Questions

1. The concept of "counting by weighing" as described in your textbook is often a difficult one for students to appreciate. Describe in your own terms what is meant by this idea.

    _____

    _____

    _____

    _____

    _____

    _____

    _____

2. Suppose a particular screw sold at the local hardware store has a mass of 2.42 g. Suppose a customer came into the store and wanted to purchase 125 such screws: how could the hardware store clerk measure out the screws without having to count them individually? Suppose a shipment of such screws arrived: if the box of screws were marked "net contents 3 lbs," how many screws would the shipment contain? Show your calculation.

3.   What do we mean by a weighted average when computing the average atomic molar mass of an element from the individual masses of the isotopes of the element?

_____

_____

_____

_____

_____

_____

_____

4.   In this experiment, it will be necessary to separate pennies manufactured before 1982 from pennies manufactured after 1982, and to exclude pennies manufactured in 1982. Why?

_____

_____

_____

_____

_____

_____

_____

_____

Name: _____ Section: _____

Lab Instructor: _____ Date: _____

EXPERIMENT 13

Counting by Weighing

Results/Observations

1. **Counting by Weighing**

   a. *Pre-1982 Pennies*

Number	Mass	Number	Mass	Number	Mass
_____	_____	_____	_____	_____	_____
_____	_____	_____	_____	_____	_____
_____	_____	_____	_____	_____	_____
_____	_____	_____	_____	_____	_____
_____	_____	_____	_____	_____	_____
_____	_____	_____	_____	_____	_____

   Average mass of your pre-1982 pennies _____

   Class average mass of pre-1982 pennies _____

   Mass of 55 pre-1982 pennies _____

   Number of pre-1982 pennies present in total mass of 310. g _____

   b. *post-1982 pennies*

Number	Mass	Number	Mass	Number	Mass
_____	_____	_____	_____	_____	_____
_____	_____	_____	_____	_____	_____
_____	_____	_____	_____	_____	_____
_____	_____	_____	_____	_____	_____
_____	_____	_____	_____	_____	_____
_____	_____	_____	_____	_____	_____

   Average mass of your post-1982 pennies _____

   Class average mass of post-1982 pennies _____

   Mass of 75 post-1982 pennies_____

   Number of post-1982 pennies present in total mass of 250. g_____

2.   Effect of Isotopes on Average Atomic Mass

Average mass of a penny based on average of individual masses   _____

Average mass of a penny based on "weighted average" method   _____

Questions

1.   How do the average masses of a penny reported by your classmates compare to your own average penny mass? Are there significant differences in the average masses reported, or are they all comparable? What factor(s) might account for differences in the average masses reported by your classmates?

_____

_____

_____

_____

_____

_____

_____

_____

2.   The element krypton can exist in numerous isotopic forms. Below are listed the seven most abundant isotopes of krypton, along with their exact masses and relative abundances. Based on this data, calculate the average atomic mass of a krypton atom.

Isotope	Mass	Abundance
Kr-78	77.920	0.35 %
Kr-80	79.916	2.28 %
Kr-82	81.913	11.58 %
Kr-83	82.914	11.49 %
Kr-84	83.911	57.00 %
Kr-86	85.911	17.30 %

EXPERIMENT 14

The Length of a Molecule

Objective

The approximate length of a stearic acid molecule will be determined by measuring the diameter of a film of stearic acid that forms on the surface of an aqueous medium.

Introduction

When petroleum products are accidentally spilled in a waterway, one of the problems with cleanup is the tendency for the petroleum product to spread rapidly into a thin layer on the surface of the water. This tendency results in what is commonly referred to as an "oil slick." The components of petroleum are mostly hydrocarbons and do not mix with or dissolve in water. Rather, they spread out into what eventually becomes a layer of petroleum that is a single molecule thick (a *monolayer*).

We can make use of this tendency of hydrocarbons to form a monolayer on the surface of water in determining the length of a molecule. Consider the structure of stearic acid:

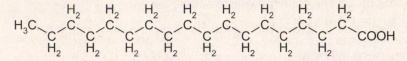

The stearic acid molecule consists of a chain of 18 carbon atoms, which makes it very much like a hydrocarbon. However, stearic acid also contains a polar carboxyl group at one end, which makes the end of the molecule able to be attracted by water molecules. If a droplet of stearic acid is applied to the surface of water, the droplet does spread out into an oil slick, but the molecules are not randomly arranged as the slick forms. Because the carboxyl groups of the stearic acid molecules are attracted by water molecules, the stearic acid molecules arrange themselves as the monolayer forms so that the carboxyl groups are facing the surface of the water, with the hydrocarbon chains of the molecules floating above. Once the monolayer has formed, the thickness of the monolayer basically represents, to a first approximation, the length of the stearic acid molecules. See Figure 14-1.

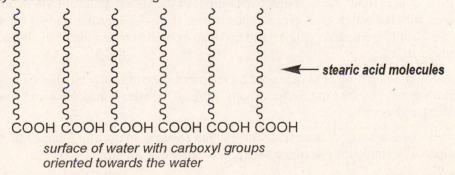

Figure 14-1. Formation of a monolayer by stearic acid on the surface of water

119

Safety Precautions	• **Safety eyewear approved by your institution must be worn at all times while you are in the laboratory, whether or not you are working on an experiment.** • **The stearic acid solution is highly flammable. No flames are permitted in the laboratory while this solution is in use. The stearic acid solution should be kept in the fume exhaust hood during the experiment.** • **Lycopodium powder can be a fire hazard if the dust gets into the air. Use the minimum amount possible to visualize the monolayer, and clean up spills of lycopodium powder with a damp sponge.**

Apparatus/Reagents Required

- 9-inch aluminum pie plate or cake pan

- glass Pasteur pipets and bulb

- lycopodium powder (or finely-powdered sulfur)

- stearic acid solution in hexane (0.15 g stearic acid per liter of solution)

- 10-mL graduated cylinder

- wash bottle with distilled water

- transparent plastic ruler

Procedure

1. Calibration of the Pipet

The stearic acid will be applied to the surface of the water using a Pasteur pipet (a glass pipet with a long, thin tip) and bulb. Take the pipet/bulb, your 10-mL graduated cylinder, watch glass and a small, clean, dry beaker to the fume exhaust hood where the stearic acid solution is available.

In the exhaust hood, obtain about 5 mL (roughly measured) of the stearic acid solution in the small beaker and keep it covered with the watch glass except when in use. Transfer approximately 3 mL of the stearic acid solution to the 10-mL graduated cylinder, and make an exact determination of the liquid level in the cylinder. Record.

To determine the volume of a droplet as delivered by the pipet, determine the number of droplets of stearic acid solution that is required to transfer exactly 1 mL of the stearic acid solution from the beaker to the graduated cylinder. Record.

Based on the number of droplets of the stearic acid solution required to transfer 1 mL of the solution, calculate the volume of a single droplet of the stearic acid solution. Record.

Repeat the procedure to obtain a second value for the volume of a droplet as delivered by your pipet. Record. Calculate the average of your two results. Record.

2. Preparation of the Monolayer

Obtain an aluminum pie plate or cake pan and a transparent plastic ruler.

Clean the plate or pan with soap and water. Rinse the plate or pan under running tap water for at least five minutes to remove all traces of soap. Finally, rinse the plate or pan with several portions of distilled water from your wash bottle. It is important that the plate or pan be clean of all oils and soap, since these substances are similar to stearic acid and will interfere with the formation of a monolayer.

Place the plate or pan on a clean, flat surface in the fume exhaust hood, and add distilled water to approximately a 1 inch depth.

To make the formation of the stearic acid monolayer more visible, you are going to apply a very fine coating of a powder to the surface of the water in the pan. Lycopodium powder is a finely divided moss that floats on water. It offers little resistance to the spread of an oil slick.

Obtain a small amount of lycopodium powder and spread it out thinly on a small sheet of paper.

From a distance of approximately 1 foot, blow the powder onto the surface of the water in the pan, using a short, sharp puff of air. If this is done carefully, the powder will settle uniformly on the surface of the water in the pan without clumping. If lycopodium powder is not available, finely-powdered sulfur may be used as an alternative.

The stearic acid solution is prepared with hexane as solvent. Hexane is a very nonpolar hydrocarbon with a high volatility. When the stearic acid solution is applied to the water surface, the hexane will rapidly evaporate, leaving only the stearic acid, which will then spread out into the monolayer.

Fill the pipet you prepared above with the stearic acid solution. Position the tip of the pipet approximately 1 cm above the surface water in the pan, trying to get as close as you can to the *center* of the pan. See Figure 14-2.

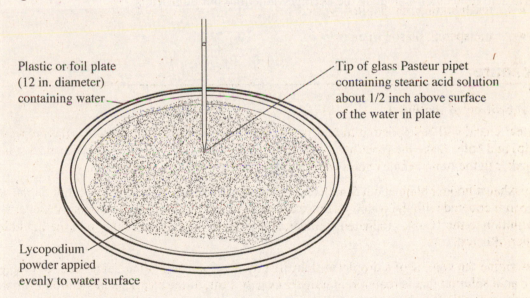

Plastic or foil plate (12 in. diameter) containing water

Tip of glass Pasteur pipet containing stearic acid solution about 1/2 inch above surface of the water in plate

Lycopodium powder appied evenly to water surface

Figure 14-2. *The plate contains water to which a dusting of lycopodium has been applied. The glass Pasteur pipet is positioned just above the water surface for applying the stearic acid solution.*

Carefully add 10 to12 drops of the stearic acid solution to the water in the pan (count them exactly). The solution may initially bead up on the surface of the water. Watch the stearic acid solution carefully: as the solvent evaporates, the monolayer will be generated and will push the lycopodium powder outward, forming a ring.

Quickly record the diameter of the monolayer formed using the transparent plastic ruler. If the ring is not a perfect circle, measure the diameter at several places and use the average of these. If there are wind currents in the laboratory, the ring sometimes elongates into an ellipse.

Clean out the plate or pan thoroughly, and repeat the procedure twice. For each run, record the exact number of droplets of the stearic acid solution used, and the diameter of the monolayer that forms.

Calculations

From the diameter of the stearic acid ring in each of your runs, calculate the area of the stearic acid monolayer for each run. For a circle, $A = \pi r^2$ where r represents the radius of the circle. The radius of a circle is half the diameter.

The stearic acid solution contained 0.15 g of stearic acid per liter. From the average volume of a droplet as delivered by your pipet (Part 1) and the number of droplets used to form the monolayer (Part 2), calculate the mass of stearic acid present in the monolayer for each of your trials.

> For example, suppose in Part 1 it was found that the volume of a droplet was 0.035 mL, and in Part 2, ten drops were used in forming the monolayer.

$$10\,drops \times \frac{0.035\,mL}{1\,drop} \times \frac{1\,L}{1000\,mL} \times \frac{0.15\,g\,stearic\,acid}{1\,L} = 5.3 \times 10^{-5}\,g\,stearic\,acid$$

Given that the density of pure stearic acid is 0.848 g/mL, calculate the volume of pure stearic acid that was present in the monolayer in each of your trials.

From the volume of pure stearic acid present in the monolayer, and from the area of the monolayer, calculate the thickness of the monolayer for each of your trials, which is effectively the length of the stearic acid molecule. Calculate the average value for your three trials.

Name: _____ Section: _____

Lab Instructor: _____ Date: _____

EXPERIMENT 14

The Length of a Molecule

Pre-Laboratory Questions

1. It is an old maxim that "oil and water don't mix." Explain why.

    _____

    _____

    _____

    _____

    _____

    _____

    _____

    _____

    _____

2. Using your textbook or a chemical encyclopedia, look up the atomic radius of carbon.

    _____

3. Using your answer to Question 2, calculate the approximate length of a linear, continuous chain of 18 carbon atoms. Show your calculation.

4.   Explain why the stearic acid molecules in this experiment are expected to orient themselves in such a way that a monolayer forms.

_____

_____

_____

_____

_____

_____

_____

_____

_____

_____

_____

5.   Describe the method to be used to determine the volume of one drop of stearic acid as delivered by the pipet used for the experiment.

_____

_____

_____

_____

_____

_____

_____

_____

_____

_____

_____

Name: _____ Section: _____

Lab Instructor: _____ Date: _____

EXPERIMENT 14

The Length of a Molecule

Results/Observations

1. Calibration of the Pipet

    Number of droplets equivalent to 1 mL Trial 1 _____ Trial 2 _____

    Average number of droplets equivalent to 1 mL (from above) _____

2. Preparation of the Monolayer

	Trial 1	*Trial 2*	*Trial 3*
Number of droplets to form monolayer	_____	_____	_____
Diameter of monolayer, cm	_____	_____	_____

3. Calculations

	Trial 1	*Trial 2*	*Trial 3*
Area of the monolayer	_____	_____	_____
Mass of stearic acid in the monolayer	_____	_____	_____
Volume of stearic acid in the monolayer	_____	_____	_____
Thickness of the monolayer	_____	_____	_____

    Average thickness of the monolayer (length of the molecule) _____

Questions

1. It was necessary for you to wash the pan or plate to be used for the experiment very thoroughly with soap to remove any oils and then to rinse thoroughly to remove all traces of the soap. Why would residual oils or soap have interfered with the experiment?

    _____

    _____

    _____

    _____

    _____

    _____

2.   In Pre-Laboratory Question 3 for this experiment, you calculated the approximate length of a linear, continuous chain of 18 carbon atoms using the atomic radius of carbon. Considering stearic acid contains a chain of 18 carbon atoms, how does the length of the stearic acid molecule you determined in this experiment compare to the length calculated in the Pre-Laboratory Question? Is your measured length shorter or longer than the calculated length? How might you account for any differences between the measured and calculated lengths?

_____

_____

_____

_____

_____

_____

_____

_____

_____

_____

_____

3.   For what purpose was *lycopodium* powder used in this experiment? Use an encyclopedia or an online resource to look up the nature of lycopodium powder and some other uses it has.

_____

_____

_____

_____

_____

_____

_____

_____

_____

_____

_____

126

EXPERIMENT 15

Stoichiometry and Limiting Reactant

Objective

The stoichiometric ratio in which hydrochloric and sulfuric acids react with sodium hydroxide will be determined, and the concept of limiting reactant will be demonstrated.

Introduction

The concept of limiting reactant is very important in the study of the stoichiometry of chemical reactions. The *limiting reactant* is the reactant that controls the amount of product possible for a process; once the limiting reactant has been consumed, no further reaction can occur.

Consider the following balanced chemical equation

$$A + B \rightarrow C$$

Suppose we set up a reaction in which we combine 1 mol of substance A and 1 mol of substance B. From the stoichiometry of the reaction, and from the amounts of A and B used, we would predict that exactly 1 mol of product C should form.

Suppose we set up a second experiment involving the same reaction, in which we again use 1 mol of A, but this time we only use 0.5 mol of B. Clearly, 1 mol of product C will *not* form. According to the balanced chemical equation, substances A and B react in a 1:1 ratio, and with only 0.5 mol of substance B, there is not enough substance B present to react with the entire 1 mol of substance A. Substance B would be the limiting reactant in this experiment. There would be 0.5 mol of unreacted substance A remaining after the reaction had been completed, and only 0.5 mol of product C would be formed.

Suppose we set up a third experiment, in which we combine 1 mol of substance A with 2 mols of substance B. In this case, substance A would be the limiting reactant (which would control how much product C is formed), and there would be an excess of substance B present once the reaction was complete.

One way of determining the extent of reaction for a chemical process is to monitor the temperature of the reaction system. Chemical reactions nearly always absorb or liberate heat energy as they occur, and the amount of heat energy transferred will be directly proportional to how much product has formed. In this experiment, you will monitor temperature changes as an indication of the extent of reaction, using a simple thermometer.

In this experiment, you will prepare solutions of hydrochloric acid and sulfuric acid, and will determine the stoichiometric ratio with which these acids react with the base sodium hydroxide. You will perform several trials of these acid-base reactions in which the amount of each reagent used is systematically varied between the trials. By monitoring the temperature changes that take place as the reaction occurs, you will have an index of the extent of reaction. The maximum extent of reaction will occur when the reactants have been mixed together in the correct stoichiometric ratio for reaction. If the reactants for a particular trial are *not* in the correct stoichiometric ratio, then one of these reactants will limit the extent of reaction, and will also limit the temperature increase observed during the experiment.

127

Safety Precautions	• Safety eyewear approved by your institution must be worn at all times while you are in the laboratory, whether or not you are working on an experiment. • The 3 *M* acid and base solutions used in this experiment may be corrosive to skin, eyes and clothing. Wear a lab coat or apron while diluting the concentrated solutions. Wash after handling. Inform the instructor if these reagents are spilled. Sulfuric acid in particular is dangerous because the acid is not volatile: if even a dilute solution is spilled, the acid will concentrate as water evaporates from it. • When diluting a concentrated acid or base solution, measure out the quantity of *water* required first. Then *slowly*, and with stirring, pour the concentrated acid or base to the water. *Always add acids to water and not the other way around!* • Be careful not to break the thermometer when stirring the solutions during reaction. If mercury is spilled, inform the instructor immediately so that the mercury can be cleaned up. Beware of broken glass from the thermometer.

Reagents/Apparatus Required

- 3.0 *M* solutions of HCl, H_2SO_4 and NaOH
- plastic foam cups
- thermometer
- ruler

Procedure

Record all data and observations directly on the report pages in ink.

Check your Pre-Laboratory calculations with the instructor before preparing the solutions needed for this experiment.

1. Reaction Between HCl and NaOH

When preparing dilute acid/base solutions, slowly add the more concentrated stock solution to the appropriate amount of water while stirring.

Using 600-mL beakers for storage, prepare 500. mL each of 1.0 *M* NaOH and 1.0 *M* HCl, using the stock 3.0 *M* solutions available. Be sure to measure the amounts of concentrated acid or base and water required with a graduated cylinder and add acid to the required amount of water slowly.

Stir the solutions vigorously with a stirring rod for one minute to mix. Keep the solutions covered with a watch glass when not in use. Frequently, diluting a concentrated solution with water will result in a temperature increase. Allow the solutions to stand for 5–10 minutes until they have come to the same temperature (within ±0.2°C). Be sure to rinse and wipe the thermometer before switching between solutions.

Obtain a plastic foam cup for use as a reaction vessel: an insulated cup is used so that the heat liberated by the reactions will not be lost to the room during the experiment.

Using a graduated cylinder, measure out 45.0 mL of 1.0 M HCl. Pour the HCl solution into the plastic cup and determine the temperature of the HCl (to the nearest 0.2°C). Record this temperature on your report sheet.

Measure 5.0 mL of 1.0 M NaOH into a 10-mL graduated cylinder.

Add the NaOH to the HCl in the plastic cup all at once and *carefully* stir the mixture with the thermometer.

Determine and record the *highest temperature reached* as the reaction occurs.

Rinse and dry the plastic cup.

Perform the additional reaction trials indicated in Table I on the report sheet. In each case, measure the amounts of each solution carefully with a graduated cylinder. For each trial, record the initial temperature of the solution in the plastic cup, as well as the highest temperature reached during the reaction.

2. Reaction Between H₂SO₄ and NaOH

When preparing dilute acid/base solutions, slowly add the more concentrated stock solution to the appropriate amount of water with stirring.

Using 600-mL beakers for storage, prepare 500. mL each of 1.0 M NaOH and 1.0 M H₂SO₄, using the stock 3.0 M solutions available. Be sure to measure the amounts of concentrated acid or base and water required with a graduated cylinder.

Stir the solutions vigorously with a stirring rod for one minute to mix them. Keep the solutions covered with a watch glass when not in use.

Allow the solutions to stand for 5–10 minutes until they have come to the same temperature (within ±0.2°C). Be sure to rinse and wipe the thermometer before switching between solutions.

Using the procedure described earlier, perform the reaction trials indicated in Table II on the report sheet for H₂SO₄ and NaOH. In each case, measure the amounts of each solution carefully with a graduated cylinder. For each trial, record the initial temperature of the solution in the plastic cup, as well as the highest temperature reached during the reaction.

Interpretation of Results

Based on the volume and concentration of the solutions used in each experiment, calculate the number of moles of each reactant used in each trial. For example, if you used 25.0 mL of 1.0 M HCl solution, the number of moles of HCl present would be

$$25.0 \text{ mL} \times \frac{1 \text{L}}{1000 \text{ mL}} \times 1.0 \text{ } M = 0.025 \text{ mol}$$

Record these values in Tables I and II on the report sheet.

To most clearly demonstrate how the extent of reaction in each of the experiments is determined by the limiting reactant, prepare two graphs of your experimental data. One graph should represent the reaction between HCl and NaOH; the second graph should represent the reaction between H₂SO₄ and NaOH.

Set up your graphs so that the vertical axis represents the temperature change measured for a given trial. Set up the horizontal axis to represent the number of moles of NaOH used in the trial. Plot each of the data points with a sharp pencil.

You should notice that each of your graphs consists of a set of ascending points and a set of descending points. For each of these sets of points, use a ruler to draw the best straight line through the points.

The intersection of the lines through the ascending points and the descending points in each graph represents the *maximum extent of reaction* for the particular experiment. From the maximum point on your graph, draw a straight line down to the horizontal axis. Read off the horizontal axis the number of moles of NaOH (and acid) that have reacted at this point.

According to your graph, in what stoichiometric ratio do HCl and NaOH react? According to your graph, in what stoichiometric ratio do H_2SO_4 and NaOH react?

For each of your graphs, indicate which points represent the acid as limiting reactant, and which points represent NaOH as limiting reactant.

Staple your two graphs to the report pages.

EXPERIMENT 15

Stoichiometry and Limiting Reactant

Pre-Laboratory Questions

1. Magnesium metal reacts with chlorine gas to produce magnesium chloride, $MgCl_2$. Write the balanced chemical equation for the reaction.

   _____

2. If 5.00 g magnesium is combined with 10.0 g of chlorine, show by calculation which substance is the limiting reactant, and calculate the theoretical yield of magnesium chloride for the reaction.

3. Show below how to calculate how many mL of 3.0 M HCl are required to prepare 500. mL of 1.0 M HCl solution.

4. Consider the following data table, which is similar to the sort of data you will collect in this experiment. Complete the entries in the table. On graph paper from the end of this manual, construct a graph of the data, plotting the temperature change measured for each run versus the number of moles of substance A used for the run. Use your graph to determine the stoichiometric ratio in which substances A and B react. Explain your reasoning.

mL 1.0 M A	mol of A	mL 1.0 M B	mol of B	temperature change, °C
5.0	_____	45.0	_____	1.7
10.0	_____	40.0	_____	3.4
15.0	_____	35.0	_____	5.2
20.0	_____	30.0	_____	6.9
25.0	_____	25.0	_____	8.5
30.0	_____	20.0	_____	6.8
35.0	_____	15.0	_____	5.1
40.0	_____	10.0	_____	3.3
45.0	_____	5.0	_____	1.8

_____

_____

_____

_____

_____

_____

_____

_____

_____

_____

EXPERIMENT 15

Stoichiometry and Limiting Reactant

Results/Observations

1. **Reaction Between HCl and NaOH**

    Table I

mL HCl	mol HCl	mL NaOH	mol NaOH	temperature change, °C
45.0	_____	5.0	_____	_____
40.0	_____	10.0	_____	_____
35.0	_____	15.0	_____	_____
30.0	_____	20.0	_____	_____
25.0	_____	25.0	_____	_____
20.0	_____	30.0	_____	_____
15.0	_____	35.0	_____	_____
10.0	_____	40.0	_____	_____
5.0	_____	45.0	_____	_____

Intersection of ascending and descending portions of graph, mol NaOH _____

Stoichiometric ratio, mol HCl/mol NaOH at intersection _____

Balanced chemical equation for the reaction between HCl and NaOH

_____

2.   **Reaction Between H_2SO_4 and NaOH**

Table II

mL H_2SO_4	mol H_2SO_4	mL NaOH	mol NaOH	temperature change, °C
45.0	_____	5.0	_____	_____
40.0	_____	10.0	_____	_____
35.0	_____	15.0	_____	_____
30.0	_____	20.0	_____	_____
25.0	_____	25.0	_____	_____
20.0	_____	30.0	_____	_____
15.0	_____	35.0	_____	_____
10.0	_____	40.0	_____	_____
5.0	_____	45.0	_____	_____

Intersection of ascending and descending portions of graph, mol NaOH _____

Stoichiometric ratio, mol H_2SO_4/mol NaOH at intersection _____

Balanced chemical equation for the reaction between H_2SO_4 and NaOH

_____

Questions

1.   Suppose a similar experiment had been done using 1.0 M H_3PO_4 solution as the acid. What would be the stoichiometric ratio, mol H_3PO_4/mol NaOH, at the intersection of the ascending and descending portions of the graph for such an experiment? Explain.

_____

_____

_____

_____

_____

2.   Explain to your friend, who has never taken chemistry, what is meant by the limiting reactant.

_____

_____

_____

_____

_____

_____

_____

EXPERIMENT 16

Percentage Composition of Magnesium Oxide

Objective

The percentage composition of magnesium oxide will be determined by reaction of a weighed sample of magnesium metal with oxygen.

Introduction

Magnesium metal is a moderately reactive elementary substance. At room temperature, magnesium reacts only very slowly with oxygen and can be kept for long periods of time without appreciable oxide buildup.

At elevated temperatures, however, magnesium will ignite in an excess of oxygen gas, burning with an intense, white flame and producing magnesium oxide.

$$2Mg(s) + O_2(g) \rightarrow 2MgO(s)$$

Because of the brightness of its flame, magnesium is used in flares and photographic flashbulbs.

In this experiment, however, you will be heating magnesium in a closed container called a *crucible*, exposing it only gradually to the oxygen of the air. Under these conditions, the magnesium will undergo a more controlled oxidation, gradually turning from shiny metal to grayish-white powdered oxide.

Because the air also contains a great deal of nitrogen gas, a portion of the magnesium being heated may be converted to magnesium nitride, Mg_3N_2, rather than to magnesium oxide. Magnesium nitride will react with water and, on careful heating, is converted into magnesium oxide.

$$Mg_3N_2(s) + 3H_2O(l) \rightarrow 3MgO(s) + 2NH_3(g)$$

The ammonia produced by this reaction can be detected by its odor, which is released on heating of the mixture.

Magnesium is a Group IIA metal, and its oxide would be expected to have the formula MgO. Based on this formula and the molar masses of Mg (24.31 g) and O (16.00 g), the theoretical percentage of magnesium in MgO should be:

$$\% \, Mg = \frac{24.31 \text{ g Mg}}{(24.31 \text{ g Mg} + 16.00 \text{ g O})} \times 100 = \frac{24.31 \text{ g Mg}}{40.31 \text{ g MgO}} \times 100 = 60.31\% \, Mg$$

In this experiment you will weigh out a small sample of magnesium, and then heat it to convert it to magnesium oxide (which will also be weighed). From the mass of magnesium taken, and the mass of oxygen gained during the reaction, you will be able to confirm this percentage composition for MgO.

Safety Precautions	• Safety eyewear approved by your institution must be worn at all times while you are in the laboratory, whether or not you are working on an experiment. • Magnesium burns with an intense, bright flame that may be damaging to the eyes. If the magnesium accidentally ignites while being heated in the crucible, immediately cover the crucible and stop heating. Do not look directly at the burning magnesium. • When water is added to the crucible, the contents of the crucible may spatter if heated too strongly. Use only gentle heating to evaporate the water. Do not heat strongly until it is certain that all water has been removed. • Use crucible tongs to handle the hot crucible and cover. Remember that the crucible, cover, ring stand, ring and clay triangle will be hot. Do not attempt to adjust or move any part of the apparatus unless it has cooled for at least 5 minutes after heating stops. • Hydrochloric acid is damaging to skin, eyes and clothing. If HCl is spilled, wash immediately and inform the instructor.

Apparatus/Reagents Required

- porcelain crucible and cover
- crucible tongs
- clay triangle
- magnesium turnings (or ribbon)
- pH paper
- 6 M HCl

Procedure

Record all data and observations directly in your notebook in ink.

Obtain a crucible and cover and examine them. The crucible and cover are extremely fragile and are very expensive. Use caution in handling them.

If there is any loose dirt in the crucible, moisten and rub it gently with a paper towel to remove the dirt. If dirt remains in the crucible, bring it to the fume exhaust hood, add 5 to 10 mL of 6 M HCl, and allow the crucible to stand for 5 minutes. Discard the HCl and rinse the crucible with water. If the crucible is not clean at this point, consult with the instructor about other cleaning techniques, or replace the crucible. After the crucible has been cleaned, use tongs or a paper towel to handle the crucible and cover to keep finger-marks and oils off it.

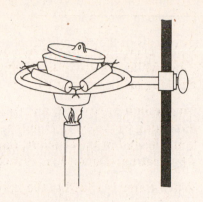

Figure 16-1.   Setup for oxidation of magnesium
The crucible is heated with its cover slightly ajar to allow oxygen in and to allow moisture to escape. Do not handle the crucible with the fingers, and allow it to cool completely before handling or weighing.

Set up a clay triangle on a ring stand. Transfer the crucible and cover to the triangle. The crucible should sit *firmly* in the triangle (the triangle's arms can be bent slightly if necessary). See Figure 16-1.

Begin heating the crucible and cover with a small flame to dry the crucible and cover. When both show no visible droplets of moisture, increase the flame to full intensity, and heat the crucible and cover for five minutes.

Remove the flame, and allow the crucible and cover to cool undisturbed on the ring stand until they return to room temperature (at least five minutes). Remember that the crucible, cover, clay triangle and iron ring are all very hot: do not attempt to move or adjust them until you have waited at least five minutes after removing the flame.

When the crucible and cover are completely cool, use tongs or a paper towel to move them to a clean, dry watch glass or flat glass plate. Do *not* place the crucible directly on the lab bench. Weigh the crucible and cover to the nearest milligram (0.001 g).

Return the crucible and cover to the clay triangle. Reheat in the full heat of the burner flame for five minutes.

Allow the crucible and cover to cool completely to room temperature on the ring stand (at least five minutes).

Reweigh the crucible after it has cooled. If the weight this time differs from the earlier weight by more than 5 mg (0.005 g), reheat the crucible for an additional five minutes and reweigh when cool. Continue the heating/weighing until the weight of the crucible and cover is constant to within 5 mg.

Add to the crucible approximately about half a teaspoon of magnesium turnings (or about 8 inches of magnesium ribbon coiled into a spiral).

Using tongs or a paper towel to protect the crucible from finger-marks, transfer the crucible/cover and magnesium to the balance and weigh to the nearest milligram (0.001 g).

Set up the crucible on the clay triangle with the cover very slightly ajar. (See Figure 16-1.) With a very small flame, begin heating the crucible gently.

If the crucible begins to smoke when heated, immediately cover the magnesium completely and remove the heat for two to three minutes. The smoke consists of the magnesium oxide product and must not be lost from the crucible.

Continue to heat gently for five to 10 minutes with the cover of the crucible slightly ajar. Remove the heat and allow the crucible to cool for one to two minutes.

Remove the cover and examine the contents of the crucible. If portions of the magnesium still demonstrate the shiny appearance of the free metal, return the cover and heat with a small flame for an additional five minutes; then re-examine the metal. Continue heating with a small flame until no shiny metallic pieces are visible.

When the shiny magnesium metal appears to have been converted fully to the dull gray oxide, return the cover to its slightly ajar position, and heat the crucible with the full heat of the burner flame for five minutes. Then slide the cover to about the half-open position, and heat the crucible in the full heat of the burner flame for an additional five minutes.

Remove the heat and allow the crucible and contents to cool completely to room temperature while still on the ring stand (at least five minutes). When the crucible is completely cool, remove the crucible from the clay triangle and set it on a sheet of clean paper on the lab bench.

With a stirring rod, gently break up any large chunks of solid in the crucible. Rinse any material that adheres to the stirring rod into the crucible with a few drops of distilled water. With a dropper, add about 10 drops of distilled water to the crucible, spreading the water evenly throughout the solid.

Return the crucible to the clay triangle, and set the cover in the slightly ajar position. With a very small flame, begin heating the crucible to drive off the water that has been added. Beware of spattering during the heating. If spattering occurs, remove the flame and close the cover of the crucible.

As the water is driven off, hold a piece of moistened pH paper (with forceps) in the steam being expelled from the crucible. Any nitrogen that had reacted with the magnesium is driven off as ammonia during the heating and should give a basic response with pH paper (you may also note the odor of ammonia).

When you are certain that all the water has been driven off, slide the cover so that it is in approximately the half-open position, and increase the size of the flame. Heat the crucible and contents in the full heat of the burner for five minutes.

Allow the crucible and contents to cool completely to room temperature undisturbed on the ring stand. When they are completely cool, weigh the crucible and contents to the nearest milligram (0.001 g).

Return the crucible to the triangle and heat for another five minutes in the full heat of the burner flame. Allow the crucible to cool completely to room temperature undisturbed on the ring stand, and then reweigh. The two measurements of the crucible and contents should give masses that agree within 5 mg (0.005 g). If this agreement is not obtained, heat the crucible for additional five-minute periods until two successive mass determinations agree within 5 mg.

If your instructor directs you to, clean out the crucible and repeat the determination.

Calculate the weight of magnesium that was taken, as well as the weight of magnesium oxide that was present after the completion of the reaction. Calculate the percentage of magnesium in the magnesium oxide from your experimental data. Calculate the mean for your two determinations.

Compare your experimentally determined percentage to the theoretical percentage calculated in the Introduction. Calculate the percent error between your percentage and the theoretical percentage.

$$\% \text{ error} = \frac{\text{experimental percentage} - \text{theoretical percentage}}{\text{theoretical percentage}} \times 100$$

Name: _____ Section: _____

Lab Instructor: _____ Date: _____

EXPERIMENT 16

Percentage Composition of Magnesium Oxide

Pre-Laboratory Questions

1. Write the balanced chemical equations for the reaction of magnesium metal with oxygen gas, and for the reaction of magnesium metal with nitrogen gas.

    _____

    _____

2. Suppose 2.145 g of magnesium is heated in air. What is the theoretical yield of magnesium oxide that should be produced? Show your calculations.

3.   Suppose for the experiment in Question 2, only 1.173 g of MgO was obtained. Calculate the percent yield for the experiment. Show your work.

4.   If heated properly in this experiment, the magnesium metal should react smoothly and rapidly. However, if the metal is not heated carefully, it may ignite. Why is it important to not look directly at the flame of burning magnesium should the metal ignite during this experiment?

_____

_____

_____

_____

_____

_____

_____

Name: _____ Section: _____

Lab Instructor: _____ Date: _____

EXPERIMENT 16

Percentage Composition of Magnesium Oxide

Results/Observations

	Trial 1	Trial 2
Mass of empty crucible (after first heating)	_____	_____
Mass of empty crucible (after second heating)	_____	_____
Mass of crucible with Mg	_____	_____
Mass of Mg taken	_____	_____
Mass of crucible/MgO (after first heating)	_____	_____
Mass of crucible/MgO (after second heating)	_____	_____
Mass of MgO in crucible	_____	_____
Mass of oxygen gained	_____	_____
% magnesium in the oxide	_____	_____

Mean % magnesium _____

Theoretical % magnesium _____

Error % _____ %

Questions

1. If magnesium oxide "smoke" had been lost during the heating of the crucible, would this have made the calculated % Mg in the product too high or too low? Explain, using a calculation to illustrate your reasoning.

_____

_____

_____

_____

_____

_____

_____

_____

Experiment 16:   Percentage Composition of Magnesium Oxide

2.   When magnesium is heated in air, most of the magnesium reacts with the oxygen of the air, but a significant amount reacts with nitrogen in the air. If water had not been added to your initial product, a *mixture* of magnesium oxide and magnesium nitride would have resulted. What error in the percentage magnesium determined for your product would have resulted if a portion of the product were magnesium nitride instead of magnesium oxide? Explain.

_____

_____

_____

_____

_____

3.   Write the balanced chemical equation showing the reaction by which magnesium nitride was converted to magnesium oxide by the addition of a few drops of water.

_____

4.   A binary compound of nitrogen and hydrogen has the following percentage composition: 82.27% nitrogen; 17.76% hydrogen. If the molar mass of the compound is determined by a separate experiment to be slightly more than 17 g, what are the empirical and molecular formulas of the compound?

_____

_____

_____

_____

_____

_____

_____

_____

_____

_____

_____

_____

142

Line Spectra: Evidence for Atomic Structure

Objective

The bright-line spectra produced when excited atoms emit electromagnetic radiation has led to our modern model for atomic structure. In this experiment you will have a chance to observe first-hand some of these spectra.

Introduction

An atom that possesses excess energy is said to be in an *excited* state. When an atom in an excited state returns to its ground state, it does so my emitting excess energy as photons of electromagnetic radiation, including visible light. The internal energy states of an atom are *discrete*; these states are of fixed, specific energies that never vary. Each type of atom (each element) has different, characteristic, discrete energy states that are different from the energy states of all other elements.

When an excited atom emits its excess energy, the energy is not emitted continuously. Atoms emit photons of only certain specific energies, which correspond exactly in energy to the changes in energy between the discrete energy states within the atom. Because atoms emit photons of only certain specific energies, the wavelengths of these emitted photons can be used to identify atoms. Typically, light from energized atoms is passed through a prism (or other device) that separates the light into its component wavelengths.

When light is separated into its component wavelengths, the pattern of different component colors produced is called a spectrum. You may have seen sunlight or light from an incandescent light bulb separated by a prism into a rainbow pattern called a continuous spectrum (see Figure 17-1 below).

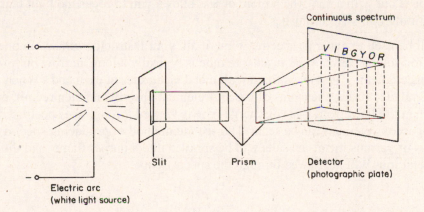

Figure 17-1. Continuous spectrum produced by a "white light" source

When light from a particular energized element is passed through a prism, a bright line spectrum is produced (see Figure 17-2 below). The colored lines occur at specific places in the spectrum (specific wavelengths) that are characteristic for each element and determined by the internal energy states in the particular atom.

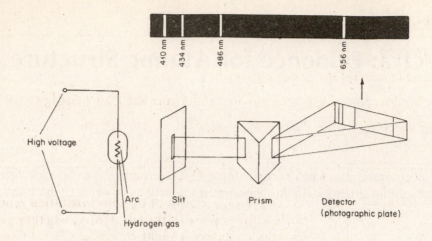

Figure 17-2. The spectrum of hydrogen
*When light from a hydrogen lamp is passed through
a prism, only certain bright lines of characteristic wavelength are observed.*

Line spectra, and more modern derivations, are routinely used in chemical analysis laboratories to detect the presence of a particular element in a sample. As an example, if an analyst believes that an unknown sample contains iron, he or she can generate a bright line spectrum for a known sample of iron, and then compare the spectrum to that of the unknown sample. If the characteristic spectral lines of iron are present in the spectrum of the unknown sample, then the sample must, indeed, contain iron. Modern instruments can also quantify spectral data, and can determine how much of a particular element is present in the sample.

In this experiment, you will use a simple spectroscope to view the bright line spectra of several elements. The spectroscope consists of a cardboard tube capped at both ends. One cap contains a circular viewing hole that has been fitted with a small piece of *diffraction grating*: a diffraction grating is a sheet of plastic that has been scored with thousands of parallel scratches per inch; it has the same effect on light as a prism (the diffraction grating separates light into its component wavelengths). The other cap on the spectroscope tube is cut with a slit, which more or less allows you to focus the light being viewed through the diffraction grating into sharper images.

Two sources will be used to generate spectra. First of all, your instructor will set up one or more gas discharge tubes for you to view. A gas discharge tube is typically a long, narrow bulb filled with a particular gaseous element. The tube is fitted with metal electrodes at each end. When a high voltage is applied to the metal electrodes, the atoms of gaseous element absorb, and then reemit, energy, partially as visible light. When the light from a gas discharge tube is viewed through the spectroscope, the bright line spectrum of the gaseous element can be seen. The second method of producing spectra you will perform yourself: samples of various metal-ion salts will be sprayed into a burner flame, and the light emitted by the energized metal ions will be viewed through the spectroscope.

Apparatus/Reagents Required

- spectroscope

- gas discharge tubes and power supply

- spray bottles containing 0.1 M solutions of the salts NaCl, KCl, $SrCl_2$, $CaCl_2$ and $CuCl_2$.

- plastic sheeting to cover the lab bench area

Safety Precautions	• **Safety eyewear approved by your institution must be worn at all times while you are in the laboratory, whether or not you are working on an experiment.** • **Gas discharge tubes also emit other wavelengths of electromagnetic radiation, in particular, ultraviolet radiation, which may be damaging to the eyes. Your safety glasses will absorb most ultraviolet radiation, so keep them *on* while observing the spectra.** • **The power supply for the gas discharge tube uses very high voltages:** *do not handle or attempt to adjust the power supply yourself.* • **The salts used in the flame tests may be toxic if ingested. Wash after handling. Wash down your lab bench to remove residues of the salts.**

Procedure

1. The spectroscope

Various types of spectroscopes may be available for your use. Examine the spectroscope you will be using for this experiment. One end of the long tube should have a round viewing hole fitted with a small piece of plastic diffraction grating. The other end of the spectroscope should have a narrow slit. Try looking at the lighting fixtures in the laboratory through the spectroscope: if your laboratory has fluorescent tube lighting, align the slit of the spectroscope parallel to the light tubes. You should see a rainbow pattern (continuous spectrum) off to either side of the slit. If you do not see such a rainbow pattern, try rotating the eyepiece of the spectroscope (it contains the diffraction grating). If you do not see a rainbow spectrum at this point, consult with the instructor.

2. Gas Discharge Tubes

Your instructor will set up a high voltage power supply and one or more gas discharge tubes. ***You should not handle this equipment yourself:*** the power supply generates more than 5,000 volts and is too dangerous to be handled by students.

The first gas discharge tube your instructor will demonstrate will be for hydrogen. After the room lights are dimmed, and the hydrogen tube has been turned on, view the hydrogen tube through the spectroscope. Align the slit of the spectroscope parallel to the length of the hydrogen tube. Rotate the eyepiece of the spectroscope until you see the bright line spectrum off to either side of the slit opening. Try to view from no more than 4 to 5 feet away from the hydrogen tube so that the spectrum is easier to see.

Record the color of the lines you see, as well as their relative order from left to right. Nearly all students will see three bright lines (red, blue-green, blue), while some students with keen eyesight may see a fourth (violet) line. Compare the spectrum you see with that shown in your textbook.

After all the students in the lab have had a chance to view the hydrogen line spectrum, your instructor will set up an oxygen gas discharge tube. View the line spectrum of oxygen and record the color of the lines you see as well as their relative order from left to right. Notice how the spectrum of oxygen is more complex than the spectrum of hydrogen. Why?

Next, your instructor will set up a gas discharge tube containing water vapor. Since water contains the elements hydrogen and oxygen, how do you expect the spectrum of water to compare to the individual spectra of hydrogen and oxygen? View the spectrum of water and record the color of the lines you see as well as their relative order from left to right.

Finally, depending on what's available in your laboratory, your instructor may set up additional gas discharge tubes for you to view. In each case, record the color and relative positions of the bright lines in the spectra.

3. Spectra from Flames

Cover your lab bench with a sheet of plastic.

Work with a partner for this part of the experiment. One person will spray solutions into the flame while the second person views the spectrum, and then the partners will switch places so that each partner gets to view all the spectra.

Set up a laboratory burner and light it. Adjust the flame so that it approximately 1 inch in height, and open the air-holes of the burner to mix oxygen with the flame.

Your instructor will provide you with several spray bottles containing metal ion solutions: $NaCl$, KCl, $SrCl_2$, $CaCl_2$ and $CuCl_2$. Since the sodium atom produces especially bright spectral lines (this is why sodium vapor is commonly used in street lighting in cities), the $NaCl$ solution should be done *after* all the other solutions have been tested: any residue of sodium on the burner may cause interference with the other elements.

While one student views the flame through the spectroscope, the other student should spray short, quick bursts of one of the metal ion solutions into the flame. The student viewing the spectrum should record the color of the lines seen, as well as their relative order from left to right. Since the solution being sprayed into the flame vaporizes quickly, several sprays may be necessary until all the spectral lines have been recorded.

The process should be repeated for each of the metal ion solutions. Allow the burner to "rest" a few minutes before switching to the next metal ion, allowing all residue of the previous solution to vaporize.

After one student has viewed and recorded all the spectra, the partners should switch places to allow the second student to also view the spectra.

After observing the spectra, rinse off the plastic sheet and wash down your lab bench top to remove residue of the salts.

Name:_____ Section:_____

Lab Instructor:_____ Date:_____

EXPERIMENT 17

Line Spectra: Evidence for Atomic Structure

Pre-Laboratory Questions

1. Use your textbook to write definitions or explanations for each of the following terms or concepts.

    a. the wavelength of electromagnetic radiation

    _____

    _____

    _____

    _____

    b. the frequency of electromagnetic radiation

    _____

    _____

    _____

    _____

    c. the speed of electromagnetic radiation

    _____

    _____

    _____

    _____

    d. the ground state of an atom

    _____

    _____

    _____

    _____

e.   why only certain wavelengths of light are emitted by excited atoms of a given element

_____

_____

_____

_____

_____

f.   what it means to say that energy levels of an atom are quantized

_____

_____

_____

_____

_____

2.   Why should you not attempt to manipulate the power supply used to illuminate the gas discharge tubes in this experiment?

_____

_____

_____

_____

_____

_____

3.   Why must safety glasses be worn while viewing the line spectra from the gas discharge tubes?

_____

_____

_____

_____

_____

Name: _____ Section: _____

Lab Instructor: _____ Date: _____

EXPERIMENT 17

Line Spectra: Evidence for Atomic Structure

Results/Observations

1. Observation of fluorescent light "rainbow" spectrum. List the colors you observed, in order from left to right, in the spectrum.

_____

_____

_____

_____

2. Gas Discharge Tubes

List the colors of the lines you observed in each spectrum, from left to right, as well as any other observations of the spectrum.

Hydrogen

_____ _____ _____ _____ _____

Oxygen

_____ _____ _____ _____ _____

Water vapor

_____ _____ _____ _____ _____

Other element(s)

_____ _____ _____ _____ _____

_____ _____ _____ _____ _____

_____ _____ _____ _____ _____

_____ _____ _____ _____ _____

Other observations on the spectra:

_____

_____

_____

3.   Spectra from Flames

List the colors of the lines you observed in each spectrum, from left to right, as well as any other observations of the spectrum.

NaCl   _____   _____   _____   _____   _____   _____

KCl   _____   _____   _____   _____   _____   _____

$SrCl_2$ _____   _____   _____   _____   _____   _____

$CaCl_2$ _____   _____   _____   _____   _____   _____

$CuCl_2$ _____   _____   _____   _____   _____   _____

Other observations on the spectra:

_____

_____

_____

_____

_____

_____

_____

Questions

1.   Some of the spectra may have contained regions that were "continuous" (that is, rather than sharp, bright lines, some regions may have shown a continuous band of color or colors). What might be responsible for such a region of continuity?

_____

_____

_____

_____

_____

_____

2.   How did the spectrum of water vapor compare to the individual spectra of hydrogen and oxygen? Why might this be expected?

_____

_____

_____

_____

_____

_____

EXPERIMENT 18

Lewis Structures and Molecular Shapes

Objective

Predicting the bonding and geometric structure for simple molecules is important in chemistry, since the properties of a molecule may be profoundly affected by these factors. In this experiment you will practice writing Lewis structures for simple molecules, and will then build three-dimensional models that demonstrate these molecules' geometric shapes. Your instructor will also have models for the structure of some common ionic solids (substances that do not exist as "molecules") for you to examine.

Introduction

The geometric shapes exhibited by molecules are often very difficult for beginning students to visualize, especially since most students' training in geometry is limited to plane (2-dimensional, or flat) geometry. Since molecules exist in three dimensions, it will be helpful for you to build some geometric models in this experiment that will allow you to see the bond angles and relative locations of the atoms in some simple molecules.

The simple model we use to predict the geometries of molecules is called the Valence Shell Electron Pair Repulsion theory, which is usually known by its abbreviation, *VSEPR*. This theory considers the environment of the most central atom in a molecule and imagines how the valence electron pairs of that central atom must be arranged in three-dimensional space around the atoms so as to minimize repulsion among the electron pairs. The general principle is: for a given number of pairs of valence electrons on the central atom, the pairs will be oriented in three-dimensional space to be as far away from each other as possible. For example, if a central atom were to have only two pairs of valence electrons around it, the electron pairs would be expected to be oriented 180° apart from each other.

The VSEPR theory also considers which valence electron pairs on the central atom are *bonding pairs* (which have atoms attached) and which are *non-bonding* (lone) *pairs*. The overall geometric shape of a molecule is determined not only by the number of valence electron pairs on the central atom, but also by which of those pairs are used for bonding to other atoms. For example, the molecules CH_4, NH_3 and H_2O are similar in that all have central atoms surrounded by four pairs of valence electrons. These molecules have different geometric shapes, however, because the relative number of bonding and lone pairs differs.

It is sometimes difficult for students to distinguish between the orientation of the electron pairs of the central atom of a molecule and the overall geometric shape of that molecule. A simple example that clearly makes this distinction concerns the case of the central atom of the molecule that has four valence electron pairs. Consider the Lewis structures of the following four molecules: hydrogen chloride, HCl; water, H_2O; ammonia, NH_3; and methane, CH_4.

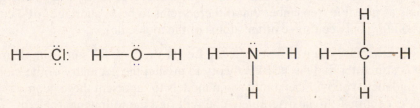

From left to right: hydrogen chloride, water, ammonia, methane

The central atom in each of these molecules is surrounded by four pairs of valence electrons. According to the VSEPR theory, these four pairs of electrons will be oriented in three-dimensional space to be as far away from each other as possible. The four pairs of electrons point to the corners of the geometric figure known as a *tetrahedron*. The four pairs of electrons are said to be *tetrahedrally oriented* and are separated by angles of approximately 109.5°.

However, three of the molecules shown are not tetrahedral in overall shape, because some of the valence electron pairs in the HCl, H_2O, and NH_3 molecules are not bonding pairs. The angular position of the bonding pairs (and hence the overall shape of the molecule) is determined by the total number of valence electron pairs on the central atom, but the non-bonding electron pairs are not included in the description of the molecules' overall shape. For example, the HCl molecule could hardly be said to be tetrahedral in shape, since there are only two atoms in the molecule. HCl is linear even though the valence electron pairs of the chlorine atom are tetrahedrally-oriented. Similarly, the H_2O molecule cannot be tetrahedral. Water is said to be *V*-shaped (bent, or nonlinear), the nonlinear shape being a result of the tetrahedral orientation of the valence electron pairs of oxygen. Ammonia's overall shape is said to be that of a trigonal (triangular) pyramid. Of the four molecules used as examples, only methane, CH_4, has both tetrahedrally-oriented valence electron pairs and an overall geometric shape that can be described as tetrahedral (since all four pairs of electrons on the central atom are bonding pairs).

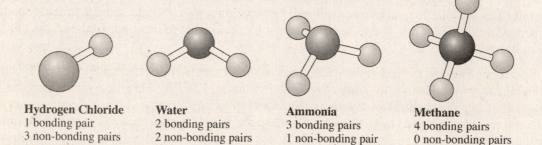

Hydrogen Chloride	**Water**	**Ammonia**	**Methane**
1 bonding pair	2 bonding pairs	3 bonding pairs	4 bonding pairs
3 non-bonding pairs	2 non-bonding pairs	1 non-bonding pair	0 non-bonding pairs

Use of the VSEPR theory to predict the geometric shape of a given molecule requires knowing how many valence electron pairs are on the most central atom of the molecule. We determine this by first drawing the Lewis electron dot structure for the molecule. The Lewis structure of a molecule is a representation that shows how the valence electrons are shared among the atoms of the molecule. Lewis structures are two-dimensional, however, and do not show a molecule's actual shape. Your textbook lists specific rules for writing Lewis structures (and gives many examples that you should review); these rules are summarized here:

1.   Determine the total number of valence electrons present in the molecule under study. For the representative elements, the number of valence electrons contributed by a given atom is indicated by its group on the periodic table.

2.   Use one pair valence electrons to form a bond between each pair of bound atoms (using either two dots or a line to represent each pair of bonding electrons).

3.   Arrange the remaining valence electrons to satisfy the duet rule for hydrogen and the octet rule for as many atoms as possible (remember that if there seems to be a "shortage" of electrons, there may be multiple bonding between some of the atoms of the molecule).

In today's laboratory exercise, you will first write the Lewis structure for some simple molecules, and will then use your Lewis structures and the VSEPR theory to predict the geometry of the molecules. With your instructor's guidance, you will then build some models to represent these geometries. You will also study some of the "exceptions" to the octet rule. You will find the following Table of Geometries helpful in your study.

Table of Geometries

Valence pairs on central atom	Arrangement of valence pairs	Bonding pairs on central atom	Geometry of molecule	Type formula
2	Linear	2	linear	AB_2
3	trigonal planar	1	linear	AB
3	trigonal planar	2	bent	AB_2
3	trigonal planar	3	trigonal planar	AB_3
4	tetrahedral	1	linear	AB
4	tetrahedral	2	bent	AB_2
4	tetrahedral	3	trigonal pyramid	AB_3
4	tetrahedral	4	tetrahedral	AB_4

The VSEPR theory helps us to predict the geometry of discrete molecules. Many common substances, however, are ionic and do not exist as molecules. Rather, such ionic substances exist as crystal lattices in which positive and negative ions effectively surround each other in a regular pattern. In a simple cubic lattice, for example, each positive ion is surrounded by six negative ions, and vice versa. The strong and extended attractive forces among all the ions in a lattice helps to explain the properties of ionic substances.

Your instructor may have available models for the crystal structures of several ionic substances (such as NaCl) and he or she will discuss the models with you. Such ionic models are very expensive, so be careful when handling them. Generally such models use different colored balls to represent different types of atoms so that you can easily visualize the arrangement of the cations and anions in the lattice.

Safety Precautions	• **Safety eyewear approved by your institution must be worn at all times while you are in the laboratory, whether or not you are working on an experiment**

Apparatus/Reagents Required

- molecular model kits
- protractors
- ionic models

Procedure

Write Lewis electron dot structures for each of the molecules listed on the report page. Use lines to indicate all pairs of bonding electrons and dots to indicate all non-bonding electrons.

Using the VSEPR theory (and the Table of Geometries), predict the geometric shape of each of the molecules for which you have drawn Lewis structures.

With your instructor's assistance, build a model to represent the geometry of each of the molecules.

Use a protractor to measure the approximate bond angles in each of your models.

Sketch a representation of each of the molecular models you have built, and indicate in your drawings the measured bond angles. Your sketches do not have to be fine artwork, but the overall shape of the molecules represented by the models must be clear. Be sure to show clearly the relative positions of all the electron pairs (both bonding and nonbonding).

Record the identity of the ionic substances whose crystal models are available in the lab on the report page.

How many negative ions surround each positive ion in the lattice? How many positive ions surround each negative ion? Sketch a simple representation of a portion of each model, illustrating the arrangement of the counter-ions.

There are several types of arrangements possible for ions in a lattice, representing the fact that the various ions may differ greatly in size, and also reflecting the stoichiometry of the substance involved. In a real ionic substance, the regular structure of the crystal lattice goes on indefinitely all the way to the edges of the crystal. Looking closely at the lattice structure, however, you will notice that there appears to be a regular arrangement of atoms that repeats over and over to generate the overall lattice. This repeating structure is sometimes called the *unit cell* for the structure. Try to determine the components of the unit cell in each of the models you examine.

EXPERIMENT 18

Lewis Structures and Molecular Shapes

Pre-Laboratory Questions

1. Summarize, in your own words, the rules for writing Lewis electron dot structures. Use an example of a simple molecule to illustrate the application of these rules.

_____

_____

_____

_____

_____

_____

_____

_____

_____

_____

_____

2. For the molecule you used as an example in Question 1 above, explain the reasoning for how the VSEPR theory would be used to deduce the geometric shape of this molecule.

_____

_____

_____

_____

_____

_____

_____

_____

_____

_____

_____

3. For each of the molecules/ions listed in the table below, draw the Lewis electron dot structure and predict the overall geometric shape of the molecule

molecule	Lewis structure	Geometric Shape
CCl_4		
NBr_3		
PO_4^{3-}		
NH_4^+		
H_2Se		
SO_4^{2-}		

Name: _____   Section: _____

Lab Instructor: _____   Date: _____

EXPERIMENT 18

Lewis Structures and Molecular Shapes

Results/Observations

Molecule	Lewis Structure	Geometric Shape	Bond Angles	Sketch
HCl				
H_2S				
NF_3				
SiH_4				
BF_3				
BeF_2				
C_2H_6				

Describe the crystal lattice structures you examined, indicating the number of positive ions surrounding each negative ion, and vice versa.

_____

_____

_____

_____

_____

Questions

1. Although you did not build any models for molecules involving multiple bonding, how would the presence of a double or triple bond help to determine the geometric shape of a molecule? Give an example of a multiple bond compound to illustrate your discussion.

    _____

    _____

    _____

    _____

    _____

    _____

    _____

2. Although models cannot show this simply, for molecules like NF_3 (whose Lewis structure you drew on the previous page), the actual bond angles may not exactly equal the angles predicted by VSEPR theory. For example, the F–N–F bond angles in NF_3 are _less_ than the 109.5° tetrahedral angle. Why do you think this might be so?

    _____

    _____

    _____

    _____

    _____

3. Use a chemical dictionary, encyclopedia or the Internet to define what chemists mean by a unit cell in a crystal lattice.

    _____

    _____

    _____

    _____

    _____

EXPERIMENT 19

Preparation and Properties of Oxygen Gas

Objective

Oxygen gas will be generated by the catalytic decomposition of hydrogen peroxide. The oxygen will be collected by displacement of water. The chemical properties of the oxygen collected will be examined.

Introduction

Elemental oxygen is essential to virtually all known forms of living creatures. Oxygen is used on the cellular level in the oxidation of carbohydrates. For example, the sugar glucose reacts with oxygen according to:

$$C_6H_{12}O_6 + 6O_2 \rightarrow 6CO_2 + 6H_2O$$

This reaction is the major source of energy in the cell (the reaction evolves energy). The chief source of oxygen gas in the earth's atmosphere is the activity of green plants. Plants contain the substance chlorophyll, which permits the reverse of the reaction to take place. Plants convert carbon dioxide and water vapor into glucose and oxygen gas. This reverse reaction is endergonic and requires the input of energy from sunlight (and is called photosynthesis).

Elemental oxygen is a colorless, odorless and practically tasteless gas. While elemental oxygen itself does not burn, it supports the combustion of other substances. Generally, a substance will burn much more vigorously in pure oxygen than in air (which is only 20% oxygen by volume). Hospitals, for example, ban smoking in patient rooms where oxygen is in use.

Oxygen gas is not very soluble in water (approximately 9 mg/L at 20°C and 1 atm), and so it may be conveniently collected by displacement of water from a container: the oxygen is bubbled into an inverted bottle filled with water that is held immersed in a larger reservoir of water. Oxygen entering the bottle pushes the water from the bottle into the reservoir. The bottle containing oxygen gas is stoppered while still under the surface of the water in the reservoir, and then the bottle is removed to the lab bench. Since oxygen gas is more dense than air, it does not escape from the bottle immediately when the bottle is subsequently opened.

Elemental oxygen is the classic oxidizing agent after which the process of oxidation was named. For example, metallic substances are oxidized by oxygen, resulting in the production of the metal oxide. Iron rusts in the presence of oxygen, resulting in the production of the rust-colored iron oxides.

$$2Fe + O_2 \rightarrow 2FeO$$

$$4Fe + 3O_2 \rightarrow 2Fe_2O_3$$

Nonmetallic substances are also oxidized by oxygen. For example, sulfur burns in oxygen, producing the two common oxides of sulfur.

$$S + O_2 \rightarrow SO_2$$

$$2S + 3O_2 \rightarrow 2SO_3$$

Oxides of sulfur in the atmosphere, produced by the burning of high-sulfur content fossil fuels, contribute to the problem of "acid rain" in parts of the United States.

159

Safety Precautions	• Safety eyewear approved by your institution must be worn at all times while you are in the laboratory, whether or not you are working on an experiment.
	• Hydrogen peroxide solution and manganese(IV) oxide may be irritating to the skin. Wash after use.
	• *Caution!* Oxygen gas supports vigorous combustion.
	• Sulfur dioxide gas is toxic and is irritating to the respiratory tract. Generate and use this gas only in the fume exhaust hood.
	• The flame of burning magnesium is intensely bright and may be damaging to the eyes. Do not look directly at the flame of burning magnesium.

Apparatus/Reagents Required

- 250-mL Erlenmeyer flask with tightly fitting two-hole stopper
- glass tubing
- thistle tube or long-stemmed funnel
- rubber tubing
- water trough
- gas collection bottles (or flasks)
- stoppers to fit the gas collection bottles (or glass plates to cover the mouth of the bottles)
- wood splints
- deflagrating spoon
- pH paper
- 3% hydrogen peroxide solution
- manganese(IV) oxide
- sulfur
- magnesium turnings
- clean iron/steel nails
- 1 *M* hydrochloric acid

Procedure

Record all data and observations directly on the report page in ink.

1. Preparation of the Gas Generator

The oxygen gas generated in this experiment will be collected by the technique of *displacement of water*. Oxygen gas is not very soluble in water, so if it is bubbled from a closed reaction system into an inverted bottle filled with water, the water will be pushed from the bottle. The gas thus collected in the bottle will be saturated with water vapor, but otherwise free of contaminating gases from the atmosphere.

Construct the gas-generating apparatus shown in Figure 19-1. The tall vertical tube shown inserted into the Erlenmeyer flask is called a *thistle tube*. It is a convenient means of delivering a liquid to the flask. If a thistle tube is not available, a long-stemmed gravity funnel may be used.

When inserting glass tubing or the stem of your thistle tube/funnel through the two-hole stopper, use plenty of glycerine to lubricate the stopper, and protect your hands with a towel in case the glass breaks.

Make certain that the two-hole stopper fits snugly in the mouth of the Erlenmeyer flask. Rinse all glycerine from the glassware before proceeding.

Set up a trough filled with water. Fill three or four gas bottles or flasks to the rim with water. Cover the mouth of the bottles/flasks with the palm of your hand and invert it into the water trough. Set aside several stoppers or glass plates; use them to cover the mouths of the bottles/flasks once they have been filled with gas.

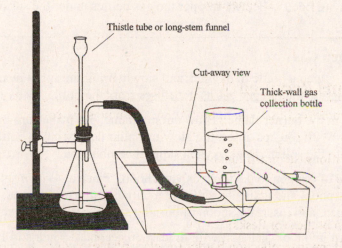

Thistle tube or long-stem funnel

Cut-away view

Thick-wall gas collection bottle

Figure 19-1. Apparatus for generation and collection of a water-insoluble gas
A long-stemmed funnel may be substituted for the thistle tube shown.

When the trough and gas collection equipment are ready, begin the evolution of gas as described in Part 2, which follows. The liquids required for the reaction are added to the gas-generating flask through the thistle tube or funnel. The liquid level in the Erlenmeyer flask must cover the bottom of the thistle tube/funnel during the generation of the gas, or gas will escape up the stem of the funnel rather than passing through the rubber tubing. Add additional liquid as needed to ensure this escape doesn't happen.

Gas should begin to bubble from the mouth of the rubber tubing as the chemical reaction begins. Allow the gas to bubble out of the rubber tubing for two to three minutes to sweep air out of the system. Add additional portions of liquid through the thistle tube/funnel as needed to continue the production of gas, and do not allow gas evolution to cease (or water may be sucked from the trough into the mixture in the Erlenmeyer flask).

After air has been swept from the system, collect the bottles of oxygen by inserting the end of the rubber tubing into the bottles of water in the trough. The oxygen gas will displace the water from the bottles. Remember to keep the mouth of the bottles under the surface of the water in the trough.

When the bottles of oxygen have been collected, stopper them while they are still under the surface of the water; and then remove them. If glass plates are used to contain the gas rather than stoppers, slide the plates under the mouths of the gas bottles while under water and remove the bottles from the water.

2. Generation and Collection of Oxygen Gas

Add approximately 1 g of manganese(IV) oxide to the Erlenmeyer flask. Add also about 15 mL of water, and shake to wet the manganese(IV) oxide (it will not dissolve). Replace the two-hole stopper (with thistle tube/funnel and glass delivery tube) in the Erlenmeyer flask, and make certain that the stopper is set tightly in the mouth of the flask. Make certain that the stem of the funnel/thistle tube extends almost to the bottom of the flask and into the liquid.

When the trough and gas collection equipment are ready, begin the evolution of oxygen by adding approximately 30 mL of 3% hydrogen peroxide to the manganese(IV) oxide (through the thistle tube or funnel). Make certain that the liquid level in the flask covers the bottom of the stem of the thistle tube/funnel or the oxygen gas will escape from the system. Add more hydrogen peroxide as needed to ensure the level is high enough.

The hydrogen peroxide should immediately begin bubbling in the flask, and oxygen gas should begin to bubble from the mouth of the rubber tubing. After sweeping air from the system for several minutes, collect five bottles of oxygen gas. Add 3 percent% hydrogen peroxide as needed to maintain the flow of gas. When the oxygen has been collected, stopper the gas bottles under the surface of the water; then remove the bottles.

3. Tests on the Oxygen Gas

Ignite a wooden splint. Open a bottle of oxygen, and slowly bring the splint near the mouth of the oxygen bottle in an attempt to ignite the oxygen gas as it diffuses from the bottle. Is oxygen itself flammable?

Ignite a wooden splint in the burner flame. Blow out the flame, but make sure that the splint still shows some glowing embers. With forceps, insert the glowing splint deeply into a bottle of oxygen gas. Explain what happens to the splint. Does oxygen support combustion?

Take a stoppered bottle of oxygen to the fume exhaust hood. Place a *very tiny* amount of sulfur in the bowl of a deflagrating spoon, and ignite the sulfur in a burner flame in the hood. Remove the stopper from the bottle of oxygen, and insert the spoon of burning sulfur into the oxygen.

Allow the sulfur to burn in the oxygen bottle for at least 30 seconds. Then add 15 to 20 mL of distilled water to the bottle, and stopper. Shake the bottle to dissolve the gases produced by the oxidation of the sulfur. Test the water in the bottle with pH paper. Write the equation for the reaction of sulfur with oxygen, and for the reaction of the oxide of sulfur with water, and explain the pH of the solution measured.

Remove the stopper from a bottle of oxygen briefly, and add 10 to 15 mL of water to the bottle. Place a *single* turning of magnesium into the bowl of a *clean* deflagrating spoon and heat until the magnesium *just* begins to burn. Open the bottle of oxygen and insert the deflagrating spoon with the burning magnesium.

When the reaction has subsided, immerse the bowl of the deflagrating spoon in the water in the bottle and shake to dislodge the oxide of magnesium. Test the water in the bottle with pH paper. Write the equation for the reaction of magnesium with oxygen, and for the reaction of magnesium oxide with water, and explain the pH of the solution measured.

Clean an iron (not steel) nail free of rust by dipping it briefly in 1 *M* hydrochloric acid; rinse with distilled water. If the surface of the nail still shows a coating of rust, repeat the cleaning process until the surface of the nail is clean.

Using tongs, heat the nail in the burner flame until it glows red. Open the fourth bottle of oxygen and drop the nail into the bottle. After the nail has cooled to room temperature, examine the nail. What is the substance produced on the surface of the nail?

EXPERIMENT 19

Preparation and Properties of Oxygen Gas

Pre-Laboratory Questions

1.   Write the balanced chemical equation for the decomposition of hydrogen peroxide, producing water and oxygen gas.

_____

2.   Use oxidation numbers to show which element is oxidized and which element is reduced in the equation you have written for Question 1.

_____

_____

_____

3.   Manganese(IV) oxide is used as a catalyst for the decomposition of hydrogen peroxide. Use your textbook or an encyclopedia of chemistry to write a definition of catalyst.

_____

_____

_____

_____

4.   Describe briefly how some gases can be collected by displacement of water. Under what circumstances could a gas not be collected by this technique?

_____

_____

_____

_____

_____

_____

_____

5.   Complete and balance the following chemical reactions, which represent the tests involving oxygen you will be performing in this experiment.

$S(s) + O_2(g) \rightarrow$

_____

$SO_2(g) + H_2O(l) \rightarrow$

_____

$Mg(s) + O_2(g) \rightarrow$

_____

$MgO(s) + H_2O(l) \rightarrow$

_____

$Fe(s) + O_2(g) \rightarrow$

_____

EXPERIMENT 19

Preparation and Properties of Oxygen Gas

Results/Observations

Observation of H_2O_2/MnO_2 Reaction

_____

_____

_____

_____

Tests on Oxygen Gas: Observations

    Attempt to ignite oxygen gas

_____

_____

    Glowing wood splint inserted in oxygen

_____

_____

    Burning of sulfur in oxygen gas

_____

_____

    Equation _____

    pH and explanation _____

    Burning of magnesium in oxygen gas

_____

_____

    Equation _____

    pH and explanation _____

    Oxidation of iron nail

    Equation _____

_____

_____

Questions

1. Why was it important that the level of liquid in the gas generator flask cover the bottom of the stem of the thistle tube or funnel?

_____

_____

_____

_____

2. The technique used in this experiment is useful for generating small quantities of oxygen gas for use in the laboratory. Use your textbook or an encyclopedia of chemistry to write the most important method for obtaining large quantities of oxygen gas for industrial processes.

_____

_____

_____

_____

_____

_____

_____

3. Oxygen is used in hospitals when patients have respiratory difficulties. Although most hospitals ban smoking completely, when oxygen is in use in a hospital room, "No smoking" signs are usually posted prominently. Explain.

_____

_____

_____

_____

4. When a gas is collected by displacement of water, the gas often appears "foggy" as it is generated. Use your textbook or an encyclopedia of chemistry to write what it means to say that a gas collected by this technique is saturated with water vapor.

_____

_____

_____

_____

_____

_____

_____

_____

EXPERIMENT 20

Charles's Law: Volume and Temperature

Objective

The gas laws describe the behavior of ideal gas samples under different pressure, volume and temperature conditions. Charles's law, which describes the relationship between a gas sample's temperature and its volume, will be investigated.

Introduction

The volume of a sample of an ideal gas is directly proportional to the absolute temperature of the gas sample (at constant pressure). The first study of the relationship between gas volumes and temperatures was made in the late eighteenth century by Jacques Charles. Charles's original statement describing his observations was that the volume of a gas sample decreased by the same factor for each degree Celsius the temperature of the sample was lowered. Specifically, Charles found that the volume of a gas sample decreased by $\frac{1}{273}$ of its volume for each degree the temperature of the sample was lowered.

The fact that the volume of a gas sample decreases in a regular way when the temperature drops led scientists to wonder what would happen if a gas sample were cooled indefinitely. If the gas sample's volume continued to decrease each time the temperature was lowered, then eventually the volume of the gas sample should be so small that lowering its temperature any further would cause the sample of gas to disappear. The temperature at which the volume of an ideal gas would be predicted to approach zero as a limit is called the *absolute zero* of temperature. Absolute zero is the lowest possible temperature. Absolute zero formed the basis for a scale of temperature (Kelvin or Absolute) that has zero as its lowest point, with all temperatures positive, relative to this temperature. The size of the degree on the Kelvin temperature scale was chosen to correspond to that of the Celsius scale. Absolute zero corresponds to -273.15 °C; it is a theoretical temperature. Real gases would never reach zero volume but rather would liquefy before reaching this temperature.

Various mathematical statements of Charles's gas law have been made. Some of these formulations are

$V = \text{constant} \times T$

$V/T = \text{constant}$

$V_1/T_1 = V_2/T_2$

Mathematically, a graph of volume *versus* temperature should be a straight line. The intercept of this line with the volume axis (i.e., $V = 0$) represents absolute zero. In this experiment, you will measure the volume of a sample of gas at several temperatures that are easily achieved in the laboratory and will plot the data obtained. By extrapolation, the intercept of this graph with the volume axis will be calculated.

The gas sample to be measured will be contained in a small glass capillary tube beneath a droplet of mercury. Because mercury is a liquid, the droplet will be free to move up and down in the capillary tube as the gas sample is heated or cooled. Because the capillary is of constant diameter, the linear height of the gas sample beneath the mercury drop can be used as a direct index of the gas's volume.

Safety Precautions	<ul><li>Safety eyewear approved by your institution must be worn at all times while you are in the laboratory, whether or not you are working on an experiment.</li><li>Mercury and its vapor are toxic. Mercury is absorbed through the skin and should not be handled.</li><li>The capillary tube containing the gas sample is extremely fragile, and if broken, the mercury in the tube may be released. If the tube is broken, inform the instructor immediately so that the mercury can be cleaned up at once.</li><li>Use tongs or a towel when handling hot beakers. Beware of burns from hot steam.</li><li>Use caution when inserting your thermometer through the rubber stopper. Use glycerine as a lubricant and protect your hands with a towel.</li></ul>

Apparatus/Reagents Required

- gas sample (air) trapped beneath a drop of mercury contained in a 3-mm O.D. capillary tube
- wooden ruler
- rubber bands
- thermometer
- ring stand, iron ring, wire gauze
- 600 mL beaker
- burner and tubing

Procedure

Record all data and observations directly in your notebook in ink.

Obtain a capillary tube containing the gas sample trapped beneath a droplet of mercury (see Figure 20-1). The capillary tube is extremely fragile and is broken easily. The capillary tube must be handled gently to avoid breaking up the mercury droplet. There must be a *single, small, unbroken droplet* of mercury in the capillary tube.

Set up a 600-mL beaker about two-thirds full of water on a ring stand in the exhaust hood and start heating the water to boiling.

Using glycerine as a lubricant, and, protecting your hands with a towel, insert the top of your thermometer through a rubber stopper so that the thermometer can still be read from 0°C to 100°C.

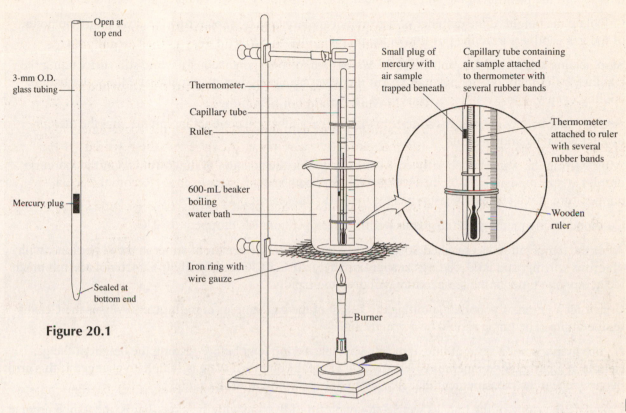

Figure 20.1

Figure 20-2

Figure 20-1. Charles's Law apparatus
The mercury droplet moves as the gas trapped beneath it is heated or cooled, giving an index of the volume of the gas sample.

Figure 20-2. Charles's Law sample attached to thermometer and ruler
Make sure the bottom of the gas sample, the bottom of the thermometer, and the zero of the ruler scale line up. Do not allow the positions of the components of the apparatus to change during the experiment.

Attach the capillary tube containing the gas sample to your thermometer with at least two rubber bands (to prevent the capillary from moving during the experiment). Align the bottom of the capillary tube with the fluid reservoir of the thermometer. See Figure 20-2.

Attach the capillary/thermometer assembly to a millimeter-scale wooden (not plastic) ruler with several rubber bands. The scale of the ruler will be used for determining the position of the mercury droplet as it moves in the capillary tube.

When the water in the beaker is boiling, clamp the thermometer/capillary/ruler assembly to the ring stand and lower it into the boiling water. Make certain that as much of the gas sample is immersed in the water bath as possible, but do not let the thermometer or gas sample touch the bottom or the walls of the beaker.

Allow the apparatus to heat in the boiling water for three to four minutes. Record the temperature of the boiling-water bath to the nearest 0.2°C. Notice that the droplet of mercury moves up in the capillary tube as the gas beneath it is heated and expands in volume.

While the gas is heating, record the position of the *inside bottom* of the glass capillary tube relative to the scale of the ruler. This measurement represents the lower end of the cylinder of gas being heated in the capillary.

169

Once the water bath reaches boiling, allow the apparatus to heat in the boiling water for several minutes, then record the position of the *bottom* of the droplet of mercury as indicated on the ruler scale.

Calculate the height of the cylinder of gas in the capillary tube by subtracting the position of the bottom of the gas capillary from the position of the mercury droplet's lower end.

Stop heating the water bath, and allow the water bath to cool spontaneously while still surrounding the capillary/ruler/thermometer apparatus. Stir the water bath periodically to make sure the water bath is cooling evenly.

As the water bath cools, the gas sample in the capillary tube will contract in volume, and the mercury droplet will move downward.

After stirring the water bath briefly each time, determine the position of the lower end of the mercury droplet at approximately 10°C intervals as the water bath and gas sample cool. Record the actual temperature at which each measurement is made (to the nearest 0.2°C).

Continue taking readings in this manner until the temperature has dropped to 30°C.

After the temperature has reached 30°C, begin adding ice to the beaker of water in small portions, with vigorous stirring, and take readings at approximately 20°C, 10°C and finally 0°C. Do not add too much ice at any one time, or the temperature will drop too rapidly.

Construct a graph of your data, plotting the height of the gas sample (in millimeters) versus the Celsius temperature. The graph should be a straight line.

If your graph is not a straight line, it means you delayed too long before reading the height of the gas sample at a particular temperature, thereby allowing the volume of the gas sample to change. If this error occurs, repeat the measurement that appears to deviate from the straight line.

Calculate the slope of the line, as well as the intercept of the line with the axis (that is, the temperature at which the volume would become zero), as directed by your instructor.

$$\text{slope} = \frac{V_2 - V_1}{T_2 - T_1}$$

$$(V_2 - V_1) = (V_2 - 0) = \text{slope}(T_2 - T_1)$$

Calculate the percent error in your determination of absolute zero.

Name: _____ Section: _____

Lab Instructor: _____ Date: _____

EXPERIMENT 20

Charles's Law: Volume and Temperature

Pre-Laboratory Questions

1. Using your textbook, summarize Charles's law for ideal gases. Give a mathematical formula that represents Charles's law.

    _____

    _____

    _____

    _____

    _____

    _____

2. The volume of a sample of ideal gas at 27 °C is 342 mL.

    a. What will the volume of the gas become if the gas is heated at constant pressure to 52°C? Show your calculations.

    b. What will the volume of the gas become if instead it is cooled to –272°C? Show your calculations.

3.   For the following volume/temperature data:

a.   Plot the data on graph paper from the back of this manual.

b.   Determine the slope of the line (see Appendix on graphing methods).

c.   Determine the intercept of the line with the volume axis (i.e., $V = 0$)

Volume, mL	Temperature, °C
83.5	30.0
86.2	40.0
88.9	50.0
91.7	60.0
94.4	70.0
97.2	80.0
100.0	90.0

Attach your graph to this page.

EXPERIMENT 20

Charles's Law: Volume and Temperature

Results/Observations

Height of gas sample   Temperature

_____   _____

_____   _____

_____   _____

_____   _____

_____   _____

_____   _____

_____   _____

_____   _____

_____   _____

_____   _____

Attach your graph to this report form.

Slope of volume versus temperature graph (indicate points used) _____

Intercept of line with axis ($V = 0$) _____

Percent error in absolute zero determination _____

Questions

1. Mercury is an environmental hazard, so it was important that the mercury be cleaned up
    immediately if any of the gas sample capillary tubes were broken during this experiment. Use a
    chemical dictionary or encyclopedia to list some of the hazards associated with mercury.

    _____

    _____

    _____

    _____

    _____

    _____

2. Occasionally, when the glass tubes containing the gas sample are prepared, water vapor is inadvertently trapped beneath the mercury plug. Why would trapped water vapor cause an error in the determination?

3. Why were you able to use the height of the gas sample in the glass tube, rather than the actual volume of the gas, in plotting the Charles's law graph?

4. In this experiment, you assumed that air behaves as an ideal gas. Under what conditions is this assumption valid?

EXPERIMENT 21

Molar Mass of a Volatile Liquid

Objective

The molar mass (molecular weight) of a volatile liquid will be determined by measuring what mass of vapor of the liquid is needed to fill a flask of known volume at a particular temperature and pressure.

Introduction

The most common instrument for the determination of molar masses in modern chemical research is the *mass spectrometer*. This instrument permits very precise determination of molar mass and also gives a great deal of structural information about the molecule being analyzed; it is of great help in the identification of new or unknown compounds.

Mass spectrometers, however, are extremely expensive and take a great deal of time and effort to calibrate and maintain. For this reason, many of the classical methods of molar mass determination are still widely applied. In this experiment, a common modification of the ideal gas law will be used to determine the molar mass of a liquid that is easily evaporated (volatile).

The ideal gas law ($PV = nRT$) indicates that the observed properties of a gas sample [pressure (P), volume (V), and temperature (T)] are directly related to the quantity of gas in the sample (n, moles). For a given container of fixed volume at a particular temperature and pressure, only one possible quantity of gas can be present in the container:

$$n = \frac{PV}{RT}$$

By careful measurement of the mass of the gas sample under study in the container, the molar mass of the gas sample can be calculated, since the molar mass, M, merely represents the number of grams of the volatile substance per mole:

$$M = \frac{g}{n}$$

In this experiment, a small amount of easily volatilized liquid will be placed in a flask of known volume. The flask will be heated in a boiling water bath and will be equilibrated with atmospheric pressure. From the volume of the flask used, the temperature of the boiling water bath, and the atmospheric pressure, the number of moles of gas contained in the flask may be calculated. From the mass of liquid required to fill the flask with vapor when it is in the boiling water bath, the molar mass of the liquid may be calculated.

A major assumption is made in this experiment that may affect your results. We assume that the vapor of the liquid behaves as an ideal gas. Actually, a vapor behaves *least* like an ideal gas under conditions similar to those in which the vapor would liquefy. The unknown liquids provided in this experiment have been chosen, however, so that the vapor will approach ideal gas behavior.

Safety Precautions	<ul><li>Safety eyewear approved by your institution must be worn at all times while you are in the laboratory, whether or not you are working on an experiment.</li><li>Assume that the vapors of your liquid unknown are toxic. Work in a fume exhaust hood or other well-ventilated area.</li><li>The liquid unknowns may be harmful to the skin, or may be absorbed through the skin. Avoid contact, and wash immediately if the liquid is spilled.</li><li>A boiling water bath is used to heat the unknown liquid, and there may be a tendency for the boiling water to splash when the flask containing the unknown liquid is immersed in it. Exercise caution.</li><li>Use tongs or a towel to protect your hands from hot glassware.</li><li>The unknown liquid samples may be environmental hazards. Do not pour down the drain. Dispose of the liquid samples as directed by the instructor.</li></ul>

Apparatus/Reagents Required

- 500-mL Erlenmeyer flask and 1,000-mL beaker
- aluminum freezer foil
- hot plate
- needle or pin
- oven (that can heat to 110°C)
- unknown liquid sample

Procedure

Record all data and observations directly in your notebook in ink.

Prepare a 500-mL Erlenmeyer flask by cleaning the flask and then drying it completely. The flask must be completely dry, since any water present will vaporize under the conditions of the experiment and will adversely affect the results. An oven may be available for heating the flask to dryness, or your instructor may describe another technique.

Cut a square of thick (freezer) aluminum foil to serve as a cover for the flask. Trim the edges of the foil so that it neatly covers the mouth of the flask but does not extend far down the neck.

Prepare a 1,000-mL beaker for use as a heating bath for the flask. The beaker must be large enough for most of the flask to be *covered* by boiling water when being heated in the boiling water bath. Add water to the beaker and test the level of water in the beaker when the flask is immersed completely in the water bath, adding or removing water as necessary. Set up the beaker on a hot plate in the exhaust hood, but do not begin to heat the water bath yet.

Weigh the dry, empty flask with its foil cover to the nearest milligram (0.001 g).

Obtain an unknown liquid and record its identification number.

Add 3 to 4 mL of liquid to the dry 500-mL Erlenmeyer flask. Cover the flask with the foil cover, making sure that the foil cover is tightly crimped around the rim of the flask. Punch a single small hole in the foil cover with a needle or pin.

Add 2–3 boiling stones, and heat the water in the 1,000-mL beaker to boiling. When the water in the beaker begins to boil, adjust the temperature of the hot plate so that the water remains boiling but does not splash from the beaker.

Immerse the flask containing the unknown liquid in the boiling water so that most of the flask is covered with the water of the heating bath (see Figure 21-1). Clamp the neck of the flask to maintain the flask in the boiling water.

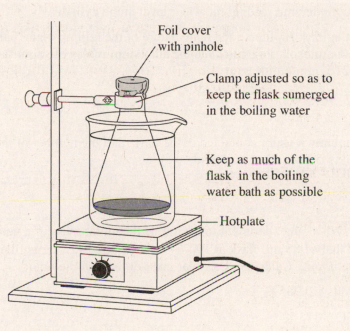

Figure 21-1. Apparatus for determination of the molar mass of a volatile liquid
Most of the flask containing the unknown liquid must be beneath the surface of the boiling-water bath.

Watch the unknown liquid carefully. The liquid will begin to evaporate rapidly, and its volume will decrease. The amount of liquid placed in the flask is much *more* than will be necessary to fill the flask with vapor at the boiling-water temperature. Excess vapor will be observed escaping through the pinhole made in the foil cover of the flask.

When it appears that all the unknown liquid has vaporized, and the flask is filled with vapor, continue to heat for one more minute. Remove the flask from the boiling-water bath, using the clamp on the neck of the flask to protect your hands from the heat.

Set the flask on the lab bench, remove the clamp (*Caution!*), and allow the flask to cool to room temperature. Liquid will reappear in the flask as the vapor in the flask cools. While the flask is cooling, measure and record the exact temperature of the boiling water in the beaker, as well as the barometric pressure in the laboratory.

When the flask has cooled completely to room temperature, carefully dry the outside of the flask to remove any droplets of water. Then weigh the flask, foil cover, and condensed vapor to the nearest milligram (0.001 g).

Repeat the determination by adding another 3- to 4-mL sample of unknown liquid. Reheat the flask until it is filled with vapor; allow the flask to cool, and then reweigh. The weight of the flask after the second sample of unknown liquid is vaporized should agree with the first determination within 0.05 g. If it does not, do a third determination.

When two acceptable determinations of the weight of vapor needed to fill the flask have been obtained, remove the foil cover from the flask and clean it out.

Fill the flask to the very rim with tap water, cover with the foil cover, and weigh the flask, the cover, and the water inside the flask to the nearest 0.1 g. Determine the temperature of the tap water in the flask. Using the density of water (calculated using the temperature of the water in the flask and the weight of water the flask contains), calculate the exact volume of the flask.

If no balance is available, the volume of the flask may be approximated by pouring the water in the flask into a 1-L graduated cylinder and reading the water level in the cylinder.

Using the volume of the flask (in liters), the temperature of the boiling-water bath (in kelvins), and the barometric pressure (in atmospheres), calculate the number of moles of vapor the flask is capable of containing. $R = 0.0821$ L atm/mol K.

$$n = \frac{PV}{RT}$$

Using the weight of unknown vapor contained in the flask, and the number of moles of vapor present, calculate the molar mass of the unknown liquid.

$$M = \frac{g}{n}$$

Example: When an experiment to determine the molar mass of a volatile liquid is performed, it is found that a 985 mL flask at 100°C and 1.02 atm pressure is filled by 1.97 g of vapor of the volatile liquid. Calculate the molar mass of the volatile liquid.

985 mL = 0.985 L

100°C = 373 K

$$n = \frac{PV}{RT} = \frac{(1.02 \text{ atm})(0.985 \text{ L})}{\left(0.0821 \dfrac{\text{L atm}}{\text{mol K}}\right)(373 \text{ K})} = 0.0328 \text{ mol}$$

$$M = \frac{g}{n} = \frac{1.97 \text{ g}}{0.0328 \text{ mol}} = 60.0 \text{ g/mol}$$

EXPERIMENT 21

Molar Mass of a Volatile Liquid

Pre-Laboratory Questions

1. The method used in this experiment is sometimes called the **vapor density method**. Beginning with the ideal gas equation, $PV = nRT$, show how the density of a vapor may be determined by this method.

    _____

    _____

    _____

    _____

    _____

    _____

    _____

2. If 2.31 g of the vapor of a volatile liquid is able to fill a 498-mL flask at 100°C and 775 mm Hg, calculate the molar mass of the liquid. Calculate the density of the vapor under these conditions. Show your work.

3.   Why is a vapor unlikely to behave as an ideal gas near the temperature at which the vapor would liquefy?

_____

_____

_____

_____

_____

_____

_____

4.   It is important that the flask used for molar mass determination is completely dry before adding the volatile liquid unknown to it. Why? What would happen to any droplets of wash-water remaining in the flask when the temperature of the flask is raised to 100°C in the boiling water bath?

_____

_____

_____

_____

_____

_____

_____

_____

Name: _____ Section:_____

Lab Instructor: _____ Date:_____

EXPERIMENT 21

Molar Mass of a Volatile Liquid

Results/Observations

Identification number of unknown liquid _____

Mass of empty flask and cover _____

	Sample 1	*Sample 2*
Mass of flask/cover/vapor	_____	_____
Temperature of vapor, °C	_____	_____
Temperature of vapor, K	_____	_____
Pressure of vapor, mm Hg	_____	_____
Pressure of vapor, atm	_____	_____
Mass of flask/cover with water, g	_____	_____
Mass of water in flask, g	_____	_____
Temperature of water in flask, °C	_____	_____
Density of water at this temperature, g/mL	_____	_____
Volume of flask	_____ mL	_____ L
Moles of vapor in flask	_____	_____
Molar mass of vapor	_____	_____
Mean value of molar mass	_____	_____

Questions

1. Two methods were described for determining the volume of the flask used for the molar mass determination. Which method will give a more precise determination of the volume? Why?

    _____

    _____

    _____

    _____

    _____

2. It was important that the flask be completely dry before the unknown liquid was added so that water present would not vaporize when the flask was heated. A typical single drop of liquid water has a volume of approximately 0.05 mL. Assuming the density of liquid water is 1.0 g/mL, how many moles of water are in one drop of liquid water, and what volume would this amount of water occupy when vaporized at 100°C and 1 atm?

    _____
    _____
    _____
    _____
    _____
    _____
    _____
    _____
    _____
    _____
    _____
    _____
    _____

3. The pressure of the vapor in this experiment when the flask was full of vapor was assumed to be equal to the pressure of the atmosphere in the laboratory. Why is this assumption valid?

    _____
    _____
    _____
    _____
    _____
    _____
    _____
    _____
    _____
    _____
    _____

EXPERIMENT 22

The Solubility of a Salt

Objective

In this experiment, you will determine the solubility of a given salt. You also will prepare the solubility curve for your salt.

Introduction

The term solubility in chemistry has both general and specific meanings. In everyday situations, we might say that a salt is "soluble," meaning that we were able to dissolve a sample of the salt in a particular solvent.

In a more specific sense, however, the solubility of a salt refers to a definite numerical quantity. Typically, the solubility of a substance is indicated as the number of grams of the substance that will dissolve in 100 g of the solvent. More often than not the solvent is water. In that case the solubility could also be indicated as the number of grams of solute that dissolve in 100 mL of water (the density of water is very near to 1.0 g/mL under most conditions).

Since solubility refers to a specific, experimentally-determined amount of a substance, it is not surprising that handbooks of chemical data and online databases contain extensive lists of solubilities of various substances. In looking at such data in a handbook or online, you will notice that the temperature at which the solubility was measured is always given. Solubility changes with temperature. For example, if you like your tea extra sweet, you have undoubtedly noticed that it is easier to dissolve two teaspoons of sugar in hot tea than in iced tea. For many substances, the solubility increases with increased temperature. For a number of other substances, however, the solubility decreases with increasing temperature.

For convenience, graphs of solubility are often used, rather than lists of solubility data. A graph of the solubility of a substance versus the temperature will clearly indicate whether or not the solubility increases or decreases as the temperature is raised. If the graph is carefully prepared, the specific numerical solubility can be read from the graph.

It is important to distinguish experimentally between whether a substance is soluble in a given solvent, and how fast (or, how easily) the substance will dissolve. Sometimes an experimenter may wrongly conclude that a salt is not soluble in a solvent when the solvent is merely dissolving at a very slow rate. The speed at which a solvent dissolves has nothing to do with the final maximum quantity of solute that can enter a given amount of solvent. In practice, we use various techniques to speed up the dissolving process; grinding the solute to a fine powder or stirring/shaking the mixture, for example. Such techniques will not affect the final amount of solute that ultimately dissolves.

The solubility of a salt in water represents the amount of solute necessary to reach a state of equilibrium between saturated solution and un-dissolved additional solute. This number is a constant for a given solute/solvent combination at a constant temperature.

Safety Precautions	• Safety eyewear approved by your institution must be worn at all times while you are in the laboratory, whether or not you are working on an experiment.
	• Use glycerine as a lubricant when inserting the thermometer through the rubber stopper. Protect your hands with a towel.
	• Some of the salts used in this experiment may be toxic or environmental hazards. Wash your hands after use. Do not pour the salts down the drain. Dispose of the salts as directed by the instructor.

Apparatus/Reagents Required

- 8-inch test tube fitted with 2-hole cork or slotted rubber stopper

- copper wire for stirring

- mortar and pestle

- thermometer

- 50-mL buret

- salt for solubility determination

Procedure

Record all data and observations directly on the report page in ink.

Obtain a salt for the solubility determination. If the salt is presented as an unknown, record the code number on the report sheet (otherwise, record the formula and name of the salt). If the salt is not finely powdered, grind it to a fine powder in a mortar.

Fit an 8-inch test tube with a 2-hole slotted rubber stopper. Protecting your hands with a towel and using glycerine as a lubricant, insert your thermometer (*Caution!*) in the slotted hole of the stopper in such a way that the thermometer can still be read from 0°C to 100°C. You may need to arrange the thermometer so that the scale can be viewed through the slot in the stopper.

Obtain a length of heavy-gauge copper wire for use in stirring the salt in the test tube. If the copper wire has not been prepared for you, form a loop in the copper wire in such a way that the loop can be placed around the thermometer when in the test tube.

Fit the copper wire through the second hole in the stopper, making sure that the hole in the rubber stopper is big enough that the wire can be easily agitated in the test tube. (See Figure 22-1.)

Place about 300 mL of water in a 400-mL beaker, and heat the water to boiling.

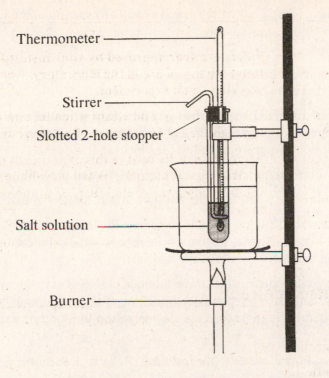

Figure 22-1. Apparatus for stirring a soluble salt
Be certain the thermometer bulb dips into the solution being determined and that the scale of the thermometer can be viewed through the slot in the stopper.

While the water is heating, weigh the empty, clean, dry, 8-inch test tube (*without* the stopper-thermometer-stirring assembly). Make the mass determination to the nearest milligram (0.001 g).

Add approximately 5 g of your salt for the solubility determination to the test tube, and reweigh the test tube and its contents. Again, make the mass determination to the nearest milligram (0.001 g).

Clean the 50-mL buret with soap and water, then rinse the buret with tap water, followed by several rinses with distilled water.

Fill the buret with distilled water. Make sure that water flows freely from the stopcock of the buret, but that the stopcock does not leak.

Record the reading of the initial water level in the buret to the nearest 0.02 mL. (Recall that water makes a meniscus. Read the bottom of the meniscus.)

In the following procedure, record on the report page each time a portion of water is added from the buret. It is essential to know the amount of water added at each point in the determination.

Add 3.00 ± 0.01 mL of water from the buret to the salt in the test tube. Record the precise amount of water added.

Attach the stopper with thermometer and stirrer, and clamp the test tube vertically in the boiling water bath.

Adjust the thermometer so that the bulb of the thermometer will be immersed in the solution in the test tube as the salt dissolves. The test tube should be set up so that the contents of the test tube are immersed fully in the boiling water. See Figure 22-1. Carefully stir the salt in the test tube until it dissolves.

185

If the salt does not dissolve completely after several minutes of stirring in the boiling water bath, remove the test tube and add 1.00 ± 0.01 mL additional water from the buret. Record. Return the test tube to the boiling water bath and stir.

If the salt is still not completely dissolved at this point, add 1.00 ± 0.01 mL water portions (one at a time) until the salt just barely dissolves. Record.

When all the salt has been dissolved, the solution will be nearly saturated, and will become saturated when the heating is stopped. Minimize the amount of time the test tube spends in the boiling water bath to restrict any possible loss of water from the test tube by evaporation.

After all the salt has dissolved completely, raise the test tube out of the boiling water.

With constant stirring, allow the solution in the test tube to cool spontaneously in the air.

Monitor the temperature of the solution carefully, and note the temperature when the first crystals of salt begin to form in the test tube: these may appear as cloudiness in the solution, or crystals may begin to form on any scratches on the interior walls of the test tube.

The first formation of crystals indicates that the solution is saturated at that temperature. Reheat the test tube in the boiling water, and make a second determination of the temperature at which the first crystal forms. If your results disagree by more than one degree, reheat the solution and make a third determination.

Add 1.00 ± 0.01 mL of additional water to the test tube. Record. Reheat the test tube in boiling water until all the solid has re-dissolved.

Remove the test tube from the boiling water and allow it to cool again spontaneously. Make a determination of the saturation temperature for solution in the same manner as indicated earlier. Repeat the determination of the new saturation temperature as a check on your measurement.

Repeat the addition of 1.00-mL water samples, with determination of the saturation temperatures, until you have at least six sets of data. Keep accurate records as to how much water has been added from the buret at each determination.

If the saturation temperature drops sharply on the addition of the 1.00 mL samples, reduce subsequent additions to 0.50 mL. If the saturation temperature does not change enough on the addition of 1.00-mL samples, increase the size of the water samples added to 2.00 mL. Keep accurate records of how much water is added.

From your data at each of the saturation temperatures, calculate the mass of salt that would have dissolved in 100 g of water at that temperature. Assume that the density of water is exactly 1.00 g/mL, so that your buret additions in milliliters will be equivalent to the mass of water being added.

On a piece of graph paper, plot the solubility curve for your salt, using *saturation temperature* on the horizontal axis and *solubility per 100 g of water* on the vertical axis. Attach the graph to your laboratory report.

Name: _____ Section: _____

Lab Instructor: _____ Date: _____

EXPERIMENT 22

The Solubility of a Salt

Pre-Laboratory Questions

1. Write a specific definition from your textbook for the following terms:

    a. saturated solution

    _____

    _____

    _____

    _____

    _____

    _____

    b. solubility

    _____

    _____

    _____

    _____

    _____

    _____

2. Why does stirring affect the rate at which a salt dissolves in water, but not the solubility of the salt in water?

    _____

    _____

    _____

    _____

    _____

    _____

    _____

    _____

3.   What does it mean to say that the solubility of a salt represents a dynamic equilibrium?

_____

_____

_____

_____

_____

_____

_____

_____

_____

4.   Suppose 3.45 g of Salt X dissolves in 8.00 mL of water at 80°C. How many grams of Salt X would dissolve in 100. mL of water at 80°C? Assume that the density of water and the solution are the same. Show your calculation.

Name: _____ Section: _____

Lab Instructor: _____ Date: _____

EXPERIMENT 22

The Solubility of a Salt

Results/Observations

mL of water used	Saturation temperature	Solubility (g solute/100 g H₂O)
_____	_____	_____
_____	_____	_____
_____	_____	_____
_____	_____	_____
_____	_____	_____
_____	_____	_____
_____	_____	_____
_____	_____	_____
_____	_____	_____
_____	_____	_____

Identity of the salt _____

Literature solubility (20°C) _____ Reference _____

Percentage error in solubility at 20°C, %_____

Attach your graph to this report page.

Questions

1. Why was it better to determine the saturation temperature while the temperature was dropping, rather than while it was rising?

    _____

    _____

    _____

    _____

    _____

    _____

    _____

2. When adding water to the salt initially, you attempted to find the minimum amount of water the salt would dissolve in at 100°C. Why was it necessary that the solution used be almost saturated?

    _____

    _____

    _____

    _____

    _____

    _____

    _____

3. The procedure indicated that the amount of time the test tube was kept in the boiling water bath should be kept as short as possible. Why was this precaution necessary?

    _____

    _____

    _____

    _____

    _____

    _____

    _____

    _____

    _____

EXPERIMENT 23

Properties of Solutions

Objective

The physical properties of a solvent are changed when a solute is dissolved in it. The magnitude of the changes that take place depend on the concentration of the solute. In this experiment, you will investigate the effect that adding a solute has on the boiling and freezing points of a solvent.

Introduction

When a solute is dissolved in a solvent, the properties of the solvent are changed by the presence of the solute. The magnitude of the change generally is proportional to the amount of solute added. Some properties of the solvent are changed only by the number of solute particles present, without regard to the particular nature of the solute. Such properties are called *colligative properties* of the solution. Colligative properties include changes in vapor pressure, boiling point, freezing point and osmotic pressure.

For example, if a nonvolatile solute is added to a volatile solvent (such as water), the amount of solvent that can escape from the surface of the liquid at a given temperature is lowered, relative to the case where only the pure solvent is present. The vapor pressure above such a solution will be lower than the vapor pressure above a sample of the pure solvent under the same conditions. Molecules of nonvolatile solute physically block the surface of the solvent, thereby preventing molecules from evaporating at a given temperature. The presence of a solute lowers the temperature at which the solution freezes (relative to the pure solvent) and raises the temperature at which the solution boils.

In general, the magnitude of the change that takes place in the solvent's properties when a solute is added depends on the number of moles of solute particles in a given amount of solvent. For the boiling point and the melting point of a solution, it is found that the change in these properties depends on the number of moles of solute particles dissolved per kilogram of solvent. The number of moles of solute dissolved per kilogram of solvent is called the *molality* of the solution, m.

$$\text{molality, } m = \frac{\text{moles of solute}}{\text{kilograms of solvent}}$$

The molality (m) of a solution (the number of moles of solute particles present per kilogram of solvent) must be carefully distinguished from the *molarity (M)* of a solution (the number of moles of solute per liter of the solution).

The decrease in freezing point (ΔT_f) when a non-volatile, non-ionizing solute is dissolved in a solvent is proportional to the molal concentration (m) of solute present in the solvent:

$$\Delta T_f = K_f m$$

K_f is a constant for a given solvent (called the "molal freezing point depression constant") and represents by how many degrees the freezing point will change when 1.00 mol of solute is dissolved per kilogram of solvent. For example, K_f for water is 1.86°C/molal, whereas K_f for the solvent benzene is 5.12°C/molal.

Similarly, the increase in the boiling point when a non-volatile, non-ionizing solute is dissolved in a solvent is given by the formula:

$$\Delta T_b = K_b m$$

K_b is a constant for a given solvent (called the "molal boiling point elevation constant"), and it represents by how many degrees the boiling point will increase when 1.00 mol of solute is dissolved per kilogram of solvent. For water, $K_b = 0.52°C/molal$.

In the preceding discussion, we considered the effect of a non-ionizing solute (such as sugar) on the freezing point of a solution. If the solute does indeed ionize in the solvent, the effect on the freezing point will be larger. The depression of the freezing point or elevation in the boiling point of a solvent is related to the number of particles of solute present in the solvent. If the solute ionizes as it dissolves, the total number of moles of all particles present in the solvent will be larger than the formal concentration indicates. For example, a 0.1 m solution of NaCl is effectively 0.1 m in both Na^+ and Cl^- ions. A 0.25 m solution of $CaCl_2$ would be 0.25 m in Ca^{2+} ion, and $2 \times 0.25 = 0.50$ m in Cl^- ion, for a total *colligative molality* of $0.25 + 0.50 = 0.75$ m.

Consider three solutions prepared by dissolving 1.00 mole of sucrose (sugar), sodium chloride and calcium chloride, respectively, in separate 1.00 kg samples of water. The freezing point depression of three solutions would be given by the following:

Sucrose $\Delta T_f = K_f m$ $\Delta T_f = (1.86\ °C/molal)(1.00\ molal) = 1.86°C$

Sodium chloride $\Delta T_f = K_f m$ $\Delta T_f = (1.86\ °C/molal)(2.00\ molal) = 3.72°C$

Calcium chloride $\Delta T_f = K_f m$ $\Delta T_f = (1.86\ °C/molal)(3.00\ molal) = 5.58°C$

Since pure water freezes at 0°C by definition, the freezing points of these three solutions would be 1.86°C, –3.72°C and –5.58°C, respectively. The larger freezing point depression of calcium chloride compared to sodium chloride is one reason calcium chloride is being used more and more as an "ice melter" for sidewalks and stairs in colder climates (calcium chloride is also much less toxic to vegetation).

Safety Precautions	• **Safety eyewear approved by your institution must be worn at all times while you are in the laboratory, whether or not you are working on an experiment.**
	• **Use glycerine as a lubricant when inserting the thermometer through the rubber stopper. Protect your hands with a towel.**
	• **Use caution when boiling water. Beware of the hot steam. Use beaker tongs or a towel to protect your hands from the heat.**

Apparatus/Reagents Required

- 100-mL beakers
- 150-mL beakers
- 250-mL beakers
- ice

- thermometer and clamp
- ring stand and ring
- wire gauze
- sucrose (sugar)
- sodium chloride
- calcium chloride
- boiling stones

Procedure

1. Freezing Point Depression

Because of the difficulties involved in trying to weigh a sample of ice accurately, we will be making only a "semi-quantitative" study of freezing point lowering.

Weigh out separately the following samples for use in the freezing point study: 34.2 g of sucrose (table sugar); 5.84 g of sodium chloride; and 11.0 g of calcium chloride. Each of these samples represents 0.1 mol of the respective substance.

Clean out three identical 100-mL glass beakers, as well as three 150-mL beakers.

Fill the 100-mL beakers with crushed ice, and pack the ice as tightly as possible in each beaker. Add 25.0 mL of distilled water to each beaker. By adding the liquid water, we are constructing the solid–liquid equilibrium system; the temperature should be exactly 0°C as long as any ice is present in the system. Although we do not know the exact mass of water (both ice and liquid) within each beaker, we can assume that the total amount in each beaker is very nearly the same.

Take the first 100-mL beaker of the ice/water mixture, and transfer it to one of the 150-mL beakers. The transfer to a larger beaker is so that you will have room to add the solute and stir the mixture.

Determine the temperature of the ice/water mixture and record. If the temperature as read on your thermometer differs from 0 °C by more than half a degree, consult with the instructor about exchanging your thermometer. Make certain that the thermometer is held in the middle of the ice/water mixture, and does not touch the walls or bottom of the beaker.

Add the weighed sample of sucrose to the 150-mL beaker containing the ice/water mixture. Stir the sugar with a stirring rod until it dissolves as much as possible. Then determine the temperature of the ice/water/sugar mixture. Record.

Repeat the process above using the other ice/water mixtures, only using the sodium chloride and calcium chloride samples in place of the sucrose. Record the freezing point of each mixture.

Assuming that we began each measurement with the same amount of solvent (water), do the freezing points you determined seem to follow the trend in freezing point depressions shown in the Introduction? Which solute produced the largest depression in the freezing point?

2. Boiling Point Elevation

Set up an apparatus for boiling as shown in Figure 23-1 using a 250-mL beaker. Do *not* add any water to the beaker at this point and do *not* begin to heat. Use a thermometer clamp or slotted rubber stopper to support the thermometer. Set up the thermometer so that the temperatures above 100°C can be easily read. Make sure the thermometer can be supported in the middle of the liquid being heated, and that it is not resting on the bottom of the beaker in contact with the burner flame.

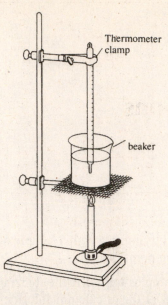

Figure 23-1. Apparatus for determination of boiling point elevation
Make certain the thermometer is freely suspended in the liquid and does not come in contact with the walls or bottom of the beaker.

Weigh out (separately) the following samples for use in the boiling point study: 34.2 g of sucrose (table sugar); 5.84 g of sodium chloride; and 11.0 g of calcium chloride. Each of these samples represents 0.1 mol of the respective substance.

With your graduated cylinder, measure out exactly 100 mL of water and transfer to the beaker that is set up for boiling point elevation determination. Add a boiling stone to the beaker. Begin heating the water and bring it to a gentle boil. The boiling stone does not dissolve in the water and will have no effect on the boiling point.

When the water is gently boiling, determine its temperature and record it. The boiling point of water varies with atmospheric pressure, but should be very near 100°C.

Protecting your hands with beaker tongs or a towel, discard the water in the beaker.

With your graduated cylinder, measure out a fresh 100-mL sample of water and add to the beaker. Add the weighed sucrose sample you prepared to the water, and stir with a stirring rod to dissolve the sucrose. Add a boiling stone.

Begin heating the sugar solution until it begins to boil very *gently*. Determine the boiling point of the solution and record.

Repeat the boiling point determination, using fresh 100-mL portions of water and using the sodium chloride and calcium chloride samples in place of the sucrose. Use a fresh boiling stone each time.

Assuming that the density of water is very near to 1.00 g/mL, you have effectively determined the boiling points of 1.00 *m* solutions of sucrose, sodium chloride, and calcium chloride (since each solid sample represented 0.1 mol, and since 100 mL of water has a mass of 0.100 kg).

Given the equation $\Delta T_b = K_b m$ for the boiling point elevation, calculate what boiling points would be expected for 1.00 molal solutions of the three solutes in this experiment. K_b for water is 0.52 °C/molal.

How do the boiling points you measured compare with those that would be predicted by the $\Delta T_b = K_b m$ equation discussed in the Introduction? How might you account for any discrepancies?

EXPERIMENT 23

Properties of Solutions

Pre-Laboratory Questions

1. Use your textbook to define the following terms

    a.   solute

    _____
    _____
    _____
    _____
    _____

    b.   solvent

    _____
    _____
    _____
    _____
    _____

    c.   solution

    _____
    _____
    _____
    _____
    _____

    d.   molality

    _____
    _____
    _____
    _____
    _____

2.   A solution is made of 1.95 g of sucrose ($C_{12}H_{22}O_{11}$) dissolved in 9.97 g of water. Given the K_f and K_b values for water listed in the Introduction, calculate the freezing and boiling points of the sucrose solution. Show your work.

3.   Consider 0.1 *molal* solutions of the following solutes in water. What would be the colligative molality of each solution? Explain.

        0.1 *molal* glucose   _____

        0.1 *molal* calcium chloride   _____

        0.1 *molal* aluminum chloride   _____

        0.1 *molal* sodium chloride   _____

EXPERIMENT 23

Properties of Solutions

Results/Observations

1. Freezing Point Depression

Solute	Observed Freezing Point, °C	Calculated Freezing Point, °C	Difference, °C
Sucrose			
Sodium chloride			
Calcium chloride			

Is your data consistent with the examples given for these substances in the Introduction? Explain.

_____

_____

_____

_____

2. Boiling Point Elevation

Solute	Observed Boiling Point, °C	Calculated Boiling Point, °C	Difference, °C
Sucrose			
Sodium chloride			
Calcium chloride			

Are the boiling points you observed for these solutions consistent with the trend you would expect based on the number of particles produced when each of the respective solvents dissolve? Explain.

_____

_____

_____

_____

_____

197

Questions

1. Describe, on a microscopic basis, how a liquid boils and also how the presence of a non-volatile solute changes boiling.

    _____
    _____
    _____
    _____
    _____
    _____

2. Freezing point depression and boiling point elevation are just two of the colligative properties of a solution. Use a scientific encyclopedia to write two additional colligative properties of a solution and tell how those properties depend on the concentration of solute in the solution.

    _____
    _____
    _____
    _____
    _____
    _____
    _____
    _____
    _____
    _____

3. Calcium chloride has a larger effect, per mole, on the freezing point of water than does sodium chloride, which is one reason calcium chloride is commonly used to melt ice on sidewalks and stairways in cold climate areas. Use the Internet or other references to find other reasons why calcium chloride is preferred for this purpose over sodium chloride.

    _____
    _____
    _____
    _____
    _____

EXPERIMENT 24

The Determination of Calcium
in Calcium Supplements

Objective

The amount of calcium ion in a dietary calcium supplement will be determined by titration with standard ethylenediaminetetraacetic acid solution (EDTA).

Introduction

The concentration of calcium ion in the blood is very important for the development and preservation of strong bones and teeth. Absorption of too much calcium may result in the build-up of calcium deposits in the joints. More commonly, a low concentration of calcium in the blood may lead to the leaching of calcium from bones and teeth. The absorption of calcium is controlled, in part, by Vitamin D.

Growing children in particular need an ample supply of calcium for the development of strong bones. Milk, which is a good natural source of calcium consumed primarily by children, is usually fortified with Vitamin D to promote calcium absorption. Similarly, outdoor recess for elementary school children is mandated by law in many states because exposure to sunshine allows the body to synthesize Vitamin D.

In later adulthood, particularly among women, the level of calcium in the blood may decrease to the point where calcium is removed from bones and teeth to replace the blood calcium, making the bones and teeth much weaker and more susceptible to fracture. This condition is called osteoporosis and can become very serious, causing shrinking of the skeleton and severe arthritis. For this reason, calcium supplements are often prescribed in an effort to maintain the proper concentration of calcium ion in the blood. Typically, such calcium supplements consist of calcium carbonate, $CaCO_3$.

A standard analysis for calcium ion involves titration with a standard solution of ethylenediaminetetraacetic acid (EDTA). In a titration experiment, a standard reagent of known concentration is added slowly to a measured volume of a sample of unknown concentration until the reaction is complete. From the concentration of the standard reagent, and from the volumes of standard and unknown taken, the concentration of the unknown sample may be calculated. In titration experiments, the sample of solution of unknown concentration is typically measured with a pipet, and the volume of standard solution required to react with the unknown sample is measured with a buret.

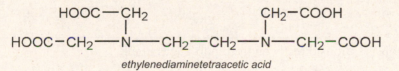

ethylenediaminetetraacetic acid

EDTA is a type of molecule called a *complexing agent*; it is able to form stable, stoichiometric (usually 1:1) compounds with many metal 2+ ions. Reactions of EDTA with metal ions are especially sensitive to pH, and typically a concentrated buffer solution is added to the sample being titrated to maintain a relatively constant pH. A suitable indicator, which will change color when the reaction is complete, is also necessary for EDTA titrations.

Safety Precautions	• Safety eyewear approved by your institution must be worn at all times while you are in the laboratory, whether or not you are working on an experiment.
	• The hydrochloric acid and the ammonia buffer solution used in this experiment are damaging to skin and eyes. Wash after handling. In form the instructor if these reagents are spilled.
	• Ammonia is a strong cardiac stimulant and respiratory irritant. Keep the ammonia buffer solution in the fume exhaust hood at all times. Add the ammonia buffer to your sample in the fume exhaust hood. Do not bring the ammonia buffer into the room until it has been added to your sample (which will dilute it).
	• Eriochrome T indicator is toxic and will stain skin and clothing. Wash after using. Clean up all spills. The indicator is a slurry and must be shaken prior to dispensing
	• Use a funnel when adding titrant to the buret, and place the top of the buret *below eye level* when filling.
	• Use a rubber safety bulb when pipeting the samples. Do *not* pipet by mouth.

Apparatus/Reagents Required

- buret and clamp
- 25-mL pipet
- 250-mL volumetric flask
- 3 M HCl
- standard 0.0500 M EDTA solution
- pH 10 concentrated ammonia buffer solution (kept in fume exhaust hood)
- Eriochrome Black T indicator
- calcium supplement tablet

Procedure

Record all data and observations directly on the report pages in ink.

1. Dissolving of the Calcium Supplement

Obtain a calcium supplement tablet. Record the calcium content of the tablet as listed on the label by the manufacturer. Place the tablet in a clean 250-mL beaker.

Obtain 25 mL of 3 M HCl solution in a graduated cylinder. Over a five-minute period, add the HCl to the beaker containing the calcium tablet in 5-mL portions, waiting between additions until all frothing of the tablet has subsided. After the last portion of HCl has been added, allow the calcium mixture to stand for an additional five minutes to complete the dissolving of the tablet. Since such tablets usually contain various binders, flavoring agents, and other inert material, the solution may *not* be entirely homogeneous: it may appear cloudy.

While the tablet is dissolving, clean a 250-mL volumetric flask with soap and tap water. Then rinse the flask with two 10-mL portions of distilled water. Fill your plastic wash bottle with distilled water.

Taking care not to lose any of the mixture, transfer the tablet solution to the volumetric flask using a small funnel. To make sure that the transfer of the calcium solution is complete, use distilled water from your plastic wash bottle to rinse the beaker that contained the calcium sample into the volumetric flask through the funnel. Repeat the rinsing of the beaker twice more.

Use a stream of distilled water from the wash bottle to thoroughly rinse any adhering calcium solution from the funnel into the volumetric flask, and then remove the funnel.

Add distilled water to the volumetric flask until the water level is approximately 1 inch below the calibration mark on the neck of the volumetric flask. Then use a medicine dropper to add distilled water until the bottom of the solution meniscus is aligned exactly with the volumetric flask's calibration mark.

Stopper the volumetric flask securely (hold your thumb over the stopper) and then invert and shake the flask 10 to 12 times to mix the contents.

2. Preparation for Titration

Set up a 50-mL buret and buret clamp on a ring stand. Check the buret for cleanliness. If the buret is clean, water should run down the inside walls in sheets, and should not bead up anywhere. If the buret is not clean enough for use, place approximately 10 mL of soap solution in the buret and scrub with a buret brush for several minutes.

Rinse the buret several times with tap water, and then check again for cleanliness by allowing water to run from the buret. If the buret is still not clean, repeat the scrubbing with soap solution. Once the buret is clean, rinse it with small portions of distilled water, allowing the water to drain through the stopcock.

Obtain a 25-mL volumetric pipet and rubber safety bulb. Using the bulb to provide suction, fill the pipet with tap water and allow the water to drain out. During the draining, check the pipet for cleanliness. If the pipet is clean, water will not bead up anywhere on the interior during draining.

If the pipet is not clean, use the rubber bulb to pipet 10 to 15 mL of soap solution. Holding your fingers over both ends of the pipet, rotate and tilt the pipet to rinse the interior with soap for two to three minutes. Allow the soap to drain from the pipet, rinse several times with tap water, and check again for cleanliness. Repeat the cleaning with soap if needed. Finally, rinse the pipet with several 5- to 10-mL portions of distilled water.

Obtain 200 mL of standard 0.0500 *M* EDTA solution in a 400-mL beaker. Keep the EDTA solution covered with a watch glass when not in use.

Transfer 5 to10 mL of the EDTA solution to the buret. Rotate and tilt the buret to rinse and coat the walls of the buret with the EDTA solution. Allow the EDTA solution to drain through the stopcock to rinse the tip of the buret.

Rinse the buret twice more with 5- to 10-mL portions of the standard EDTA solution.

After the buret has been thoroughly rinsed with EDTA, fill the buret to slightly above the zero mark with the standard EDTA.

Allow the buret to drain until the solution level is slightly below the zero mark. Read the volume of the buret (to two decimal places), estimating between the smallest scale divisions. Record this volume as the initial volume for the first titration to be performed.

Rinse four clean 250-mL Erlenmeyer flasks with small portions of distilled water. Label the flasks as samples 1, 2, 3 and 4.

Transfer approximately 125 mL of your calcium solution from the volumetric flask to a clean, dry 250-mL beaker. Using the rubber safety bulb for suction, rinse the pipet with several 5- to 10-mL portions of the calcium solution to remove any distilled water still in the pipet. Discard the rinsings.

Pipet exactly 25 mL of the calcium solution into each of the four numbered Erlenmeyer flasks.

3. Titration of the Calcium Samples

Each sample in this titration must be treated *one at a time*. Do not add the necessary reagents to a particular sample until you are ready to titrate. Because the buffer solution used to control the pH of the calcium samples during the titration contains concentrated ammonia, it will be kept stored in the fume exhaust hood. Take calcium sample 1 to the exhaust hood and add 10 mL of the ammonia buffer. Swirl to mix.

At your bench, add five drops of Eriochrome T indicator solution to calcium sample 1. The sample should be wine-red at this point; if the sample is blue, consult with the instructor.

Add EDTA from the buret to the sample a few milliliters at a time, with swirling after each addition, and carefully watch the color of the sample. The color change of Eriochrome T is from red to blue, but the sample will pass through a gray transitional color as the endpoint is neared. You are titrating to determine at what point the red color has disappeared completely.

As the sample becomes gray in color, begin adding EDTA *dropwise* until the pure blue color of the endpoint is reached. Record the final volume used to titrate sample 1 to two decimal places, estimating between the smallest scale divisions on the buret.

Refill the buret, and repeat the titration procedure for the three remaining calcium samples. Remember not to add ammonia buffer or indicator until you are actually ready to titrate a particular sample.

4. Calculations

For each of the four titrations, use the volume of EDTA required to reach the endpoint and the concentration of the standard EDTA solution to calculate how many moles of EDTA were used. Record.

$$\text{moles EDTA used} = (\text{volume used to titrate in liters}) \times (\text{molarity})$$

Based on the fact that the calcium/EDTA reaction is of 1:1 stoichiometry, calculate how many moles of calcium ion were present in each sample titrated. Record.

Using the atomic mass of calcium, calculate the mass of calcium ion present in each of the four samples titrated, and the average mass of calcium present.

Based on the fact that the average mass of calcium calculated above represents a 25-mL sample, taken from a total volume of 250 mL used to dissolve the tablet, calculate the mass of calcium present in the original tablet.

Compare the mass of calcium present in the tablet calculated from the titration results with the nominal mass reported on the label by the manufacturer. Calculate the percent difference between these values.

$$\% \text{ difference} = \frac{\text{experimental mass } - \text{ label mass}}{\text{label mass}} \times 100$$

Name: _____ Section: _____

Lab Instructor: _____ Date: _____

EXPERIMENT 24

The Determination of Calcium in Calcium Supplements

Pre-Laboratory Questions

1. Using your textbook or an encyclopedia of chemistry, discuss some uses of calcium in the human body.

    _____
    _____
    _____
    _____
    _____
    _____
    _____

2. The reagent used as titrant in this experiment, EDTA, is a complexing agent for calcium and for many other metal ions. Use your textbook or an encyclopedia of chemistry to write a definition of this term.

    _____
    _____
    _____
    _____
    _____
    _____
    _____

3. Use your textbook or an encyclopedia of chemistry to find two other common uses of EDTA in addition to the determination of the amount of a metal ion present in a sample.

    _____
    _____
    _____
    _____

4.   The pH of the samples in this experiment will be adjusted by the use of a concentrated buffered solution. Use your textbook or an encyclopedia of chemistry to write what is meant by a buffered solution and give an example of such a solution. How does a buffered solution adjust and maintain the pH of a sample?

_____

_____

_____

_____

_____

_____

_____

_____

_____

_____

_____

_____

_____

_____

5.   Why must the pH 10 ammonia buffered solution be stored and dispensed in the fume exhaust hood?

_____

_____

_____

_____

_____

_____

_____

_____

_____

_____

Name: _____ Section: _____

Lab Instructor: _____ Date: _____

EXPERIMENT 24

The Determination of Calcium
in Calcium Supplements

Results/Observations

Dissolving of the Calcium Supplement

Observation

_____

_____

Did the tablet dissolve completely? _____

Was the solution of the tablet homogeneous? _____

Titrations

Concentration of standard EDTA solution used, M _____

Volume of calcium solution taken for titration, mL _____

Sample	1	2	3	4
final volume EDTA, mL				
initial volume EDTA, mL				
volume EDTA used, mL				
moles EDTA used				
moles Ca^{2+} ion present				
mass of Ca^{2+} present, g				

Average mass of Ca^{2+} present, g _____

Average mass of Ca^{2+} present in original tablet, g _____

Listed mass of Ca^{2+} present in tablet (from label) _____

Percent difference between experimental mass and listed mass _____

205

Questions

1. Most calcium supplements consist of *calcium carbonate*, $CaCO_3$, since this substance is readily available and relatively cheap. Based on your experimentally determined average mass of calcium ion, calculate the mass of calcium carbonate that would be equivalent to this amount of calcium ion.

   _____

   _____

   _____

   _____

   _____

2. Suggest at least two reasons why your experimentally determined amount of calcium might differ from that listed by the manufacturer of the calcium supplement you used.

   _____

   _____

   _____

   _____

   _____

   _____

   _____

   _____

3. Use your textbook or an encyclopedia of chemistry to write at least three natural sources of calcium ion that should be included in the diet.

   _____

   _____

   _____

   _____

   _____

   _____

   _____

EXPERIMENT 25

Stresses Applied to Equilibrium Systems

Objective

Many chemical reactions do not go to completion. Rather, they reach a point of chemical equilibrium before the reactants are fully converted to products. At the point of equilibrium, the concentrations of all reactants remain constant with time. In this experiment, you will investigate how outside forces acting on a system at equilibrium provoke changes within the system (Le Châtelier's principle).

Introduction

Early in the study of chemical reactions, it was noted that many chemical reactions do not produce as much product as might be expected, based on the amounts of reactants taken originally. These reactions appeared to have stopped before the reaction was complete. Closer examination of these systems (after the reaction had seemed to stop) indicated that there were still significant amounts of all the original reactants present. Quite naturally, chemists wondered why the reaction had seemed to stop, when all the necessary ingredients for further reaction were still present.

Some reactions appear to stop because the products produced by the original reaction themselves begin to react, in the reverse direction to the original process. As the concentration of products begins to build up, product molecules will react more and more frequently. Eventually, as the speed of the forward reaction decreases while the speed of the reverse reaction increases, the forward and reverse processes will be going on at exactly the same rate. Once the forward and reverse rates of reaction are identical, there can be no further net change in the concentrations of any of the species in the system. At this point, a dynamic state of equilibrium has been reached. The original reaction is still taking place but is opposed by the reverse of the original reaction also taking place. In this experiment, you will study changes made in a system already in equilibrium, by reference to Le Châtelier's principle.

Le Châtelier's principle states that if we disturb a system that is already in equilibrium, then the system will react so as to minimize the effect of the disturbance. We see the principle demonstrated in cases where additional reagent is added to a system in equilibrium, or when one of the reagents is removed from the system in equilibrium.

1. Solubility Equilibria

Suppose we have a solution that has been saturated with a solute: the solution has already dissolved as much solute as possible. If we try to dissolve additional solute, no more will dissolve because the saturated solution is in equilibrium with the solute:

$$\text{Solute} + \text{Solvent} \rightarrow \text{Solution}$$

Le Châtelier's principle is most easily seen when an ionic solute is used: Suppose we have a saturated solution of sodium chloride, NaCl. We can describe the equilibrium that exists in this way:

$$NaCl(s) \rightleftharpoons Na^+(aq) + Cl^-(aq)$$

Suppose we then try adding an additional amount of one of the ions involved in the equilibrium: For example, suppose we added several drops of concentrated HCl solution (which contains the chloride ion at high concentration). According to Le Châtelier's principle, the equilibrium would shift so as to consume some of the added chloride ion. This shift would result in a net decrease in the amount of NaCl that could dissolve. If we watched the saturated NaCl solution as the HCl was added, we should see some of the NaCl precipitate as a solid.

2. Complex Ion Equilibria

Oftentimes, dissolved metal ions will react with certain substances to produce brightly colored species called *complex ions*. For example, iron(III) ion (Fe^{3+}) reacts with the thiocyanate ion (SCN^-) to produce a dark red complex ion:

$$Fe^{3+} + SCN^- \rightleftharpoons [FeNCS^{2+}]$$

This is an equilibrium process that is easy to study because we can monitor the bright red color of $[FeNCS^{2+}]$ as an indication of the position of the equilibrium: If the solution is very red, there is a lot of $[FeNCS^{2+}]$ present; if the solution is not very red, then there must be very little $[FeNCS^{2+}]$ present.

According to Le Châtelier's principle, we can try adding additional Fe^{3+} or additional SCN^- to see the effect on the red color. We will also add a reagent (silver ion) that removes SCN^- from the system to see what effect this has on the red color, and NaOH, which precipitates Fe^{3+} as iron(III) hydroxide and removes iron from the system.

Copper(II) ion and ammonia form an intensely dark blue complex ion

$$Cu^{2+}(aq) + 4NH_3(aq) \rightleftharpoons [Cu(NH_3)_4^{2+}(aq)]$$

After first setting up this equilibrium by combining copper(II) sulfate solution and concentrated ammonia solution, we will then effectively remove NH_3 from the equilibrium system by adding hydrochloric acid solution. Ammonia and hydrogen ion from the HCl solution react with each other to form the ammonium ion, NH_4^+, which is not capable of complexing the copper(II) ion:

$$NH_3(aq) + H^+(aq) \rightleftharpoons NH_4^+(aq)$$

3. Acid/Base Equilibria

Many acids and bases exist in solution in equilibrium sorts of conditions, particularly with weak acids and bases. For example, the weak base ammonia is involved in an equilibrium in aqueous solution

$$NH_3(aq) + H^+(aq) \rightleftharpoons NH_4^+(aq)$$

Once again, we can use Le Châtelier's principle to play around with this equilibrium. We will try adding more ammonium ion or hydrogen ion to see what happens. Since none of the components of this system is itself colored, we will be adding an acid/base indicator that changes color with pH, so that we have an index of the position of the ammonia equilibrium.

The indicator we will use is phenolphthalein, which is pink in basic solution and colorless in acidic solution.

$$HIn(aq, \text{colorless}) \rightleftharpoons H^+(aq) + In^-(aq, \text{pink})$$

Phenolphthalein (represented above by In) is a weak acid in aqueous solution, in which the colorless (HIn) and colored (In^-) forms are in equilibrium. Changes in the ammonia system are reflected by a shift in the phenolphthalein equilibrium.

Another easily studied equilibrium that is dependent on the acidity of the system occurs between chromate ion, CrO_4^{2-}, and dichromate ion, $Cr_2O_7^{2-}$. If a strong acid is added to a solution of potassium chromate, the intensely yellow chromate ion is converted to the bright orange dichromate ion:

$$2CrO_4^{2-}(aq) + 2H^+(aq) \rightleftharpoons Cr_2O_7^{2-}(aq) + H_2O$$

Similarly, if a strong base is added to a solution containing the orange dichromate ion, the equilibrium is shifted in favor of the yellow chromate ion:

$$Cr_2O_7^{2-}(aq) + 2OH^-(aq) \rightleftharpoons 2CrO_4^{2-}(aq) + H_2O$$

Beginning students often think these two reactions represent oxidation-reduction processes (chromate and dichromate often *are* encountered in redox reactions); a careful examination of oxidation numbers, however, will show that there is no transfer of electrons involved in either situation.

Oleic acid ($C_{18}H_{34}O_2$) is a weak, long-chain organic acid found in fats and oils. It is not very soluble in water. The effect of added strong acid and strong base on the ionization equilibrium of oleic acid will be investigated.

Safety Precautions	• Safety eyewear approved by your institution must be worn at all times while you are in the laboratory, whether or not you are working on an experiment.
	• Concentrated ammonia is a strong respiratory and cardiac stimulant. Use concentrated ammonia only in the fume exhaust hood.
	• Concentrated hydrochloric acid is severely damaging to skin, eyes and clothing, and its vapor is highly toxic and irritating. Use concentrated HCl only in the exhaust hood, and handle the bottle with a towel or gloves to protect your hands. If HCl is spilled on the skin, wash immediately and inform the instructor. Wear gloves when using the concentrated HCl.
	• Iron(III) chloride, copper(II) sulfate and potassium thiocyanate may be toxic if ingested. Wash after use.
	• Chromium compounds (dichromate and chromate ions) are highly toxic, may burn skin and eyes, and are suspected mutagens/carcinogens. Wear gloves during their use. Inform the instructor of any spills.
	• Some of the solutions used in this experiment are environmental hazards. Do not pour down the drain. Dispose of all reagents as directed by the instructor.

Apparatus/Reagents Required

- saturated sodium chloride solution
- 12 M HCl (*Caution!*): Wear gloves when using this reagent
- 6 M HCl solution
- 0.1 M FeCl$_3$ solution
- 0.1 M KSCN solution
- concentrated ammonia solution (*Caution!*)

- ammonium chloride

- 0.1 M AgNO$_3$ solution

- phenolphthalein indicator

- 6 M NaOH solution

- 0.1 M solutions of potassium chromate and potassium dichromate

- 0.1 M barium chloride solution

- 0.5 M copper(II) sulfate solution

- oleic acid

- disposable gloves

Procedure

Record all data and observations directly on the report pages in ink.

1. Solubility Equilibria

Wear gloves when performing this section.

Obtain 2 mL of saturated sodium chloride solution in a small test tube. This solution was prepared by adding solid NaCl to water until no more would dissolve. Then the clear solution was filtered from any undissolved solid NaCl.

Add 10 drops of 12 M HCl (*Caution!*) to the NaCl solution. A small amount of solid NaCl should form and precipitate out of the solution. The crystals may form slowly, and may be very small. Examine the test tube carefully.

On the lab report sheet, describe what happens in terms of Le Châtelier's principle. What is the "stress" applied? In which direction does the equilibrium shift?

2. Complex Ion Equilibria

a. The iron/thiocyanate equilibrium

Prepare a stock sample of the bright red complex ion [FeNCS^{2+}] by mixing 2 mL of 0.1 M iron(III) chloride and 2 mL of 0.1 M KSCN solutions. The color of this mixture is too intense to use, so dilute this mixture with 100 mL of water.

Pour about 5 mL of the diluted red stock solution into each of five test tubes. Label the test tubes as 1, 2, 3, 4 and 5.

Test tube 1 will have no change made in it so that you can use it to compare its color with what will be happening in the other test tubes.

To test tube 2, add dropwise about 1 mL of 0.1 M FeCl$_3$ solution. Look carefully for any change in the intensity of the red color as the solutions are mixed. The change will occur as the drops enter the solution, but the change will dissipate with time.

To test tube 3, add dropwise about 1 mL of 0.1 M KSCN solution. Look carefully for any change in the intensity of the red color as the solutions are mixed. The change will occur as the drops enter the solution, but the change will dissipate with time.

To test tube 4, add AgNO$_3$ solution dropwise until a change becomes evident. Ag$^+$ ion removes SCN$^-$ ion from solution as a solid (silver thiocyanate).

$$Ag^+(aq) + SCN^-(aq) \rightarrow AgSCN(s)$$

To test tube 5, add 6 M NaOH until a change is evident. What is the precipitate that forms? Why did the red color of the iron/thiocyanate complex fade when NaOH was added?

Describe the intensification or fading of the red color in each test tube in terms of Le Châtelier's principle. What is the "stress" applied in each case? In which direction does the equilibrium shift?

b. The copper/ammonia equilibrium

Place 1 mL of 0.5 M copper(II) sulfate solution in a medium-size test tube. Note the color of the solution.

In the fume exhaust hood, add concentrated ammonia solution dropwise to the copper(II) sulfate solution.

Rap the test tube with your index finger after each drop of ammonia has been added so as to mix the reagents.

Initially, a light blue precipitate of copper(II) hydroxide [$Cu(OH)_2$] will form, but as more ammonia is added, the precipitate will dissolve and the dark blue copper/ammonia complex ion will form

$$Cu^{2+}(aq) + 4NH_3(aq) \rightleftharpoons [Cu(NH_3)_4^{2+}(aq)]$$

Continue adding ammonia dropwise and rapping the test tube with your index finger until all the light-blue $Cu(OH)_2$ has dissolved and only a homogeneous solution of the dark blue complex remains.

Transfer about 1 mL of the copper/ammonia solution to a clean test tube. Add 6 M HCl dropwise to the solution. Rap the test tube with your index finger after each drop of HCl has been added to mix the reagents. Continue adding 6 M HCl dropwise until a color change in the sample is evident.

Describe the color change of the solution when HCl is added in terms of Le Châtelier's principle. What stress is being applied to the copper/ammonia equilibrium? In which direction does the copper/ammonia equilibrium shift?

3. Acid/Base Equilibria

a. ammonia/phenolphthalein

Wear gloves when performing this section.

In the exhaust hood, prepare a dilute ammonia solution by adding 2 drops of concentrated ammonia (*Caution!*) to 25 mL of water.

Add three drops of phenolphthalein to the dilute ammonia solution, which will turn pink (ammonia is a weak base, and phenolphthalein is pink in basic solution).

Place about 5 mL of the pink dilute ammonia solution into each of two test tubes.

To one of the test tubes, add several small crystals of ammonium chloride (which contains the ammonium ion, NH_4^+). What happens to the color of the solution?

To the other test tube, add a few drops of 12 M HCl (*Caution!*). What happens to the color of the solution?

Describe what happens to the pink color in terms of how Le Châtelier's principle is affecting the ammonia and phenolphthalein equilibria. What is the "stress" applied in each situation? In which direction do the equilibria shift?

b. chromate/dichromate

Wear gloves when performing this section.

Place five drops of 0.1 M potassium chromate in each of two semimicro test tubes.

To one sample of chromate ion, add one drop of 12 M HCl and stir. To the second test tube, add one drop of 6 M NaOH and mix. Account for any color changes that occur in terms of Le Châtelier's principle. What is the "stress" applied? In which direction does the equilibrium shift?

Place five drops of 0.1 M potassium dichromate in each of two semi-micro test tubes.

To one sample of dichromate ion, add 1 drop of 12 M HCl and stir. To the second test tube, add one drop of 6 M NaOH and mix. Account for any color changes that occur in terms of Le Châtelier's principle. What is the "stress" applied? In which direction does the equilibrium shift?

Dispose of the chromium compounds as directed by the instructor. Do not pour down the drain.

Place five drops of 0.1 M potassium chromate in a semi-micro test tube. Add two drops of 0.1 M barium chloride solution. What happens? Barium chromate is highly insoluble in water. Add two drops of 12 M HCl to the test tube and mix. What happens? Is barium dichromate more soluble in water than is barium chromate? Explain.

c. oleic acid

Wear gloves when performing this section.

Oleic acid is a weak organic acid that is not very soluble in water. When oleic acid is placed in water, most of the oleic acid will not mix with the water, but the portion that does dissolve ionizes in a manner similar to other weak acids:

$$HA \rightleftharpoons H^+(aq) + A^-(aq)$$

Place about 10 drops of water in a semi-micro test tube, and add one drop of oleic acid. Mix and allow to settle. Make a note of the approximate thickness of the oleic acid layer in the test tube.

Add two drops of 12 M HCl to the test tube and mix. What happens to the thickness of the oleic acid layer? Explain your observation in terms of the effect of Le Châtelier's principle on the ionization equilibrium for oleic acid.

Add six to eight drops of 6 M NaOH to the test tube and mix. What happens to the thickness of the oleic acid layer? Explain your observation in terms of the effect of Le Châtelier's principle on the ionization equilibrium for oleic acid.

Name: _____ Section: _____

Lab Instructor: _____ Date: _____

EXPERIMENT 25

Stresses Applied to Equilibrium Systems

Pre-Laboratory Questions

1. Le Châtelier's principle is an important concept that enables us to manipulate equilibrium so as to achieve the maximum production of product. Explain this important principle in your own words.

_____

_____

_____

_____

_____

_____

_____

_____

_____

_____

_____

2. Perhaps the most studied equilibrium system is that by which gaseous ammonia, NH_3, is produced from elemental hydrogen and nitrogen gases. Write the balanced chemical equation for this reaction.

_____

_____

3. Suppose the ammonia synthetic reaction in Question 1 has reached a state of equilibrium at a particular temperature. Tell what effect on the net amount of ammonia obtained each of the following disturbances to the equilibrium system will have, compared to a system in which no change is made.

   a.   additional nitrogen gas is pumped into the system

   _____

   _____

   _____

b.   ammonia gas is liquefied and removed from the system as it forms

_____

_____

_____

c.   the reaction system is compressed to a smaller total volume

_____

_____

_____

d.   a very efficient catalyst is found for the reaction

_____

_____

_____

4.   Explain why a saturated solution of a salt in water represents a dynamic equilibrium. What are the opposing processes going on? How do we know the solution has reached equilibrium?

_____

_____

_____

_____

_____

_____

_____

_____

_____

_____

_____

EXPERIMENT 25

Stresses Applied to Equilibrium Systems

Results/Observations

1. Solubility Equilibria

    Effect of adding HCl to saturated NaCl observation

    _____

    Explanation

    _____

2. Complex Ion Equilibria

    a. *iron/thiocyanate equilibrium*

        Effect of adding additional Fe^{3+} to $[FeNCS^{2+}]$ observation

        _____

        Explanation

        _____

        _____

        Effect of adding additional SCN– to [FeNCS2+] observation

        _____

        Explanation

        _____

        _____

        Effect of adding Ag+ to [FeNCS2+] observation

        _____

        Explanation

        _____

        _____

        Effect of adding NaOH to [FeNCS2+] observation

        _____

        Explanation

        _____

        _____

b.   *copper/ammonia equilibrium*

Preparation of the $Cu(NH_3)_4^{2+}$ complex ion observation

_____

_____

Addition of 6 *M* HCl to the $Cu(NH_3)_4^{2+}$ observation

_____

_____

Explanation

_____

_____

_____

3.   Acid/Base Equilibria

a.   *ammonia/phenolphthalein*

Effect of adding NH_4^+ observation

_____

Explanation

_____

_____

Effect of adding HCl observation

_____

Explanation

_____

_____

b.   *chromate/dichromate*

potassium chromate

Effect of adding HCl observation

_____

Explanation

_____

_____

Effect of adding NaOH observation

_____

Explanation

_____

_____

potassium dichromate

Effect of adding HCl observation

_____

Explanation

_____

_____

Effect of adding NaOH observation

_____

Explanation

_____

_____

Effect of adding barium chloride observation

_____

Effect of adding HCl observation

_____

Explanation

_____

_____

c.   *oleic acid*

Effect of adding HCl to oleic acid observation

_____

Effect of adding NaOH to oleic acid observation

_____

Explanation

_____

_____

Questions

1.   Explain how the solubility of a salt represents a situation of dynamic equilibrium. What substances/species are in equilibrium?

_____

_____

_____

_____

_____

2.   Students often think that the reactions of dichromate and chromate ion discussed in this experiment are oxidation-reduction reactions (chromate and dichromate often are involved in redox). Show, by determining the oxidation states of the reactants and products, that there is no transfer of electrons in these reactions.

_____

_____

_____

_____

_____

_____

_____

3.   Oleic acid is a long chain (18 carbon atoms) organic acid called a "fatty acid." You found that it did not dissolve very well in water, but did dissolve readily in sodium hydroxide.

$$HA(l) + NaOH(aq) \rightarrow Na^+A^-(aq) + H_2O(l)$$

The sodium salts of fatty acids (such as sodium oleate) are given a special name and have very useful properties. Use your textbook or an encyclopedia of chemistry to describe what is special about such compounds.

_____

_____

_____

_____

_____

_____

_____

_____

_____

EXPERIMENT 26

Acid-Base Titrations

Objective

A vinegar solution of unknown concentration will be analyzed by the process of titration, using a standard sodium hydroxide solution. The sodium hydroxide solution to be used for the analysis will be prepared approximately and will then be standardized against a weighed sample of a known acidic salt.

Introduction

The technique of titration finds many applications, but is especially useful in the analysis of acidic and basic substances. Titration involves measuring the exact volume of a solution of known concentration that is required to react with a measured volume of a solution of unknown concentration, or with a weighed sample of unknown solid. A solution of accurately known concentration is called a standard solution. Typically, to be considered a standard solution, the concentration of the solute in the solution must be known to four significant figures.

In many cases (especially with solid solutes) it is possible to prepare a standard solution by accurate weighing of the solute, followed by precise dilution to an exactly known volume in a volumetric flask. Such a standard is said to have been prepared *determinately*. One of the most common standard solutions used in analyses, however, cannot be prepared in this manner.

Solutions of sodium hydroxide are commonly used in titration analyses of samples containing an acidic solute. Although sodium hydroxide is a solid, it is not possible to prepare standard sodium hydroxide solutions by weight. Solid sodium hydroxide is usually of questionable purity. Sodium hydroxide reacts with carbon dioxide from the atmosphere and is also capable of reacting with the glass of the container in which it is provided. For these reasons, sodium hydroxide solutions are generally prepared to be approximately a given concentration. They are then standardized by titration of a weighed sample of a primary standard acidic substance. By measuring how many milliliters of the approximately prepared sodium hydroxide are necessary to react completely with a weighed sample of a known primary standard acidic substance, the concentration of the sodium hydroxide solution can be calculated. Once prepared, however, the concentration of a sodium hydroxide solution will change with time (for the same reasons outlined earlier). As a consequence, sodium hydroxide solutions must be used relatively quickly.

In titration analyses, there must be some way of knowing when enough titrant has been added to react exactly and completely with the sample being titrated. In an acid-base titration analysis, there should be an abrupt change in pH when the reaction is complete. For example, if the sample being titrated is an acid, then the titrant to be used will be basic (probably sodium hydroxide). When one excess drop of titrant is added (beyond that needed to react with the acidic sample), the solution being titrated will suddenly become basic. There are various natural and synthetic dyes, called indicators that exist in different colored forms at different pH values. A suitable indicator can be chosen that will change color at a pH value consistent with the point at which the titration reaction is complete. The indicator to be used in this experiment is phenolphthalein, which is colorless in acidic solutions, but changes to a pink form at basic pH.

Vinegar is a dilute solution of acetic acid, CH_3COOH. Commercially, vinegar is most commonly prepared by the fermentation of apple cider; other vinegars, such as wine vinegar or fruit-flavored vinegars, may be prepared from other fruit juices. In order to be useful and effective in food preparation

219

and other household uses, a vinegar solution should contain between 3% and 6% acetic acid by volume. Although instrumental methods are available, titration of random samples of vinegar taken from the production line is the most effective method of ensuring that the acetic acid level is within the correct range.

Safety Precautions	• Safety eyewear approved by your institution must be worn at all times while you are in the laboratory, whether or not you are working on an experiment.
	• The primary standard acidic substance, potassium hydrogen phthalate (KHP), will be kept stored in an oven to keep moisture from adhering to the crystals. Use tongs or a towel to protect your hands when removing the KHP from the oven. KHP dust may be irritating to the skin and respiratory tract. Wash after use. Avoid stirring up the dust when weighing.
	• Sodium hydroxide is extremely caustic, and sodium hydroxide dust is very irritating to the respiratory system. Do not handle the pellets of NaOH with the fingers. Wash hands after weighing the pellets. Work in a ventilated area and avoid breathing the NaOH dust.
	• Use a rubber safety bulb when pipeting. Never pipet by mouth.
	• Use a funnel when adding titrant to the buret, and place the top of the buret *below eye level* when filling.

Apparatus/Reagents Required

- buret and clamp
- buret brush
- 5-mL pipet and safety bulb
- soap
- 1-L glass or plastic bottle with stopper
- sodium hydroxide pellets
- primary standard grade potassium hydrogen phthalate (KHP)
- phenolphthalein indicator solution
- unknown vinegar sample

Procedure

Record all data and observations directly on the report pages in ink.

1. Preparation of the Burets and Pipet

For precise quantitative work, volumetric glassware must be scrupulously clean. Water should run down the inside of burets and pipets in sheets and should not bead up anywhere on the interior of the glassware.

Rinse the buret and the pipet with distilled water to see if they are clean.

If the glassware is not clean, partially fill with a few milliliters of soap solution and rotate the buret/pipet so that all surfaces come in contact with the soap.

Rinse with tap water, followed by several portions of distilled water. If the buret is still not clean, it should be scrubbed with a buret brush. If the pipet cannot be cleaned, it should be exchanged.

In the subsequent procedure, it is important that water from rinsing a pipet/buret does not contaminate the solutions to be used in the glassware. This rinse water would change the concentration of the glassware's contents. Before using a pipet/buret in the following procedures, rinse the pipet/buret with several small portions of the solution that is to be used in the pipet/buret. Discard the rinsings.

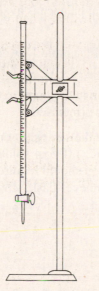

Figure 26-1. Setup for titration
The buret must be scrupulously clean, such that water runs down the interior walls in sheets and does not bead up anywhere. A funnel should be used when filling the buret, and the buret should be below eye level when the titrant solution is added to it.

2. Preparation of the Sodium Hydroxide Solution

Clean and rinse a 1-L bottle and stopper/cap. Label the bottle "Approx. 0.1 M NaOH." Put about 500 mL of distilled water into the bottle.

Weigh out approximately 4 g (0.1 mol) of sodium hydroxide pellets (*Caution!*) and transfer to the 1-L bottle using a powder (wide-stem) funnel. Stopper/cap tightly and swirl the bottle to dissolve the sodium hydroxide.

When the sodium hydroxide pellets have dissolved, add additional distilled water to the bottle until the water level is approximately 1 inch from the top. Stopper/cap and swirl thoroughly to mix.

This sodium hydroxide solution is the titrant for the analyses to follow. Keep the bottle tightly stoppered/capped when not actually in use (to avoid exposure of the NaOH to the air).

Set up a buret in the buret clamp. See Figure 26-1. Rinse and fill the buret with the sodium hydroxide solution just prepared.

3. Standardization of the Sodium Hydroxide Solution

Clean and dry a small beaker. Take the beaker to the oven that contains the primary standard grade potassium hydrogen phthalate (KHP).

Using tongs or a towel to protect your hands, remove the bottle of KHP from the oven, and pour a few grams of KHP into the beaker. If you pour too much, do *not* return the KHP to the bottle. Return the bottle of KHP to the oven, and take the beaker containing KHP back to your lab bench. Cover the beaker with a watch glass.

Allow the KHP to cool to room temperature. While the KHP is cooling, clean three 250-mL Erlenmeyer flasks with soap and water. Rinse the Erlenmeyer flasks with several small portions of distilled water.

Label the Erlenmeyer flasks as 1, 2 and 3.

When the KHP has completely cooled, weigh three samples of KHP between 0.6 and 0.8 g, one for each of the Erlenmeyer flasks. Record the exact weight of each KHP sample at least to the nearest milligram, preferably to the nearest 0.1 mg (if an analytical balance is available). Be certain not to confuse the samples.

At your lab bench, add 100 mL of water to KHP sample 1. Add two or three drops of phenolphthalein indicator solution. Swirl for several minutes to dissolve the KHP sample completely.

Record the initial reading of the NaOH solution in the buret to the nearest 0.02 mL, remembering to read across the bottom of the curved solution surface (meniscus).

Begin adding NaOH solution from the buret to the sample in the Erlenmeyer flask, swirling the flask constantly during the addition. (See Figure 26-2.) If your solution was prepared correctly, and if your KHP samples are of the correct size, the titration should require at least 20 mL of NaOH solution. As the NaOH solution enters the solution in the Erlenmeyer flask, streaks of red or pink will be visible. They will fade as the flask is swirled.

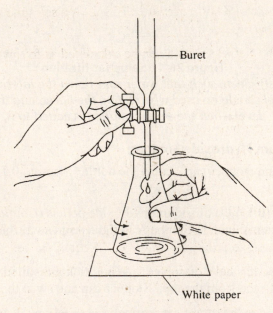

Figure 26-2. Titration technique
A right-handed person should titrate with the left hand, swirling the flask with the right hand. Titrate until one single drop causes a permanent pale pink color to appear.

Eventually the red streaks will persist for a longer and longer period of time, indicating you are approaching the endpoint of the titration.

Begin adding NaOH one drop at a time, with constant swirling, until one single drop of NaOH causes a permanent pale pink color that does not fade on swirling. A sheet of white paper under the flask makes the color easier to visualize. Record the reading of the buret to the nearest 0.02 mL.

Repeat the titration of the remaining two KHP samples. Record both initial and final readings of the buret to the nearest 0.02 mL.

Given that the molar mass of potassium hydrogen phthalate is 204.2 g, calculate the number of moles of KHP in samples 1, 2 and 3.

$$\text{moles KHP} = (\text{mass of KHP, g}) \times \frac{1 \text{ mol KHP}}{204.2 \text{ g}}$$

From the number of moles of KHP present in each sample, and from the volume of NaOH solution used to titrate the sample, calculate the concentration of NaOH in the titrant solution in moles per liter. The reaction between NaOH and KHP is of 1:1 stoichiometry. Thus at the endpoint of the titration:

moles NaOH used at the endpoint = moles KHP in the sample

The molarity of the NaOH solution is then:

$$\text{molarity of NaOH} = \frac{\text{moles of NaOH}}{\text{Liters of solution used for titration}}$$

If your three values for the concentration differ by more than 1%, weigh out an additional sample of KHP and repeat the titration. Use the average concentration of the NaOH solution for subsequent calculations for the unknown.

Example:

Suppose a 0.651 g sample of KHP required 26.25 mL of a sodium hydroxide solution to reach the phenolphthalein endpoint.

The concentration of the NaOH solution can be calculated as follows

$$0.651 \text{ g KHP} \times \frac{1 \text{ mol KHP}}{204.2 \text{ g KHP}} = 0.003188 \text{ mol KHP}$$

At the endpoint of the titration, mol NaOH = mol KHP

$$\text{Molarity of NaOH} = \frac{0.003188 \text{ mol NaOH}}{0.02625 \text{ L NaOH}} = 0.121 \, M$$

4. Analysis of the Vinegar Solution

Clean and dry a small beaker, and obtain 25 to 30 mL of the unknown vinegar solution. Cover the vinegar solution with a watch glass to prevent evaporation. Record the code number of the sample.

Clean three Erlenmeyer flasks, and label as samples 1, 2 and 3. Rinse the flasks with small portions of distilled water.

Using the rubber safety bulb to provide suction, rinse the 5-mL pipet with small portions of the vinegar solution and discard the rinsings.

Using the rubber safety bulb, pipet a 5-mL sample of the vinegar solution into each of the Erlenmeyer flasks. Add approximately 100 mL of distilled water to each flask, as well as two or three drops of phenolphthalein indicator solution.

Refill the buret with the NaOH solution and record the initial reading of the buret to the nearest 0.02 mL.

Titrate Sample 1 of vinegar in the same manner as in the standardization until one drop of NaOH causes the appearance of the pale pink color.

Record the final reading of the buret to the nearest 0.02 mL.

Repeat the titration for the other two vinegar samples.

Based on the volume of vinegar sample taken, and on the volume and average concentration of NaOH titrant used, calculate the concentration of the vinegar solution in moles per liter.

moles NaOH = (volume of NaOH used in L) × (molarity of NaOH)

moles CH_3COOH = moles NaOH from stoichiometry of reaction

$$\text{molarity of } CH_3COOH = \frac{\text{moles of } CH_3COOH}{\text{volume of sample in liters}}$$

> **Example:** Suppose a 5.00 mL sample of vinegar requires 19.31 mL of 0.121 M NaOH to reach the phenolphthalein endpoint
>
> $$\text{mol NaOH} = 0.01931 \text{ L NaOH sol.} \times \frac{0.121 \text{ mol NaOH}}{1 \text{ L NaOH sol.}} = 0.002336 \text{ mol NaOH}$$
>
> At the endpoint, mol acetic acid = mol NaOH
>
> $$\text{Molarity of acetic acid} = \frac{0.002336 \text{ mol acetic acid}}{0.00500 \text{ L acetic acid taken}} = 0.467 \ M$$

Given that the molar mass of acetic acid is 60.0 g, and that the density of the vinegar solution is 1.01 g/mL, calculate the percent by weight of acetic acid in the vinegar solution.

Name: _____ Section:_____

Lab Instructor: _____ Date:_____

EXPERIMENT 26

Acid-Base Titrations

Pre-Laboratory Questions

1. Using a handbook, look up the formula and structure of potassium hydrogen phthalate (KHP) used to standardize the solution of NaOH in this experiment. Calculate the molar mass of KHP.

    _____

    _____

    _____

    _____

    _____

    _____

2. Using your textbook or a chemical dictionary, write the definition of an indicator.

    _____

    _____

    _____

    _____

    _____

    _____

3. What indicator will be used in this experiment? Why is this indicator suitable?

    _____

    _____

    _____

    _____

    _____

    _____

225

4.   Suppose a sodium hydroxide solution is to be standardized against pure, solid, primary-standard grade KHP. If 0.4538 g of KHP requires 44.12 mL of the sodium hydroxide solution to reach a phenolphthalein endpoint, what is the molarity of the NaOH solution? Show your calculations.

5.   If 36.32 mL of the NaOH solution described in question 4 was required to titrate a 5.00 mL sample of vinegar, calculate the molarity of acetic acid in the vinegar. Show your calculations.

EXPERIMENT 26

Acid-Base Titrations

Results/Observations

1. Standardization of NaOH Titrant

	Sample 1	Sample 2	Sample 3
Mass of KHP taken, g	_____	_____	_____
Initial NaOH reading, mL	_____	_____	_____
Final NaOH reading, mL	_____	_____	_____
Volume NaOH used, mL	_____	_____	_____
Moles of KHP present	_____	_____	_____
Molarity of NaOH, M	_____	_____	_____

Average molarity of NaOH solution _____

2. Analysis of Vinegar Solution

Identification number of vinegar sample used _____

	Sample 1	Sample 2	Sample 3
Volume vinegar taken	_____	_____	_____
Initial NaOH reading, mL	_____	_____	_____
Final NaOH reading, mL	_____	_____	_____
Volume NaOH used, mL	_____	_____	_____
Molarity of vinegar, M	_____	_____	_____

Average molarity of vinegar solution _____

% by mass acetic acid present _____

Questions

1. Commercial vinegar is generally $5.0 \pm 0.5\%$ acetic acid by mass. Assuming this composition in your unknown, by how much were you in error in your analysis?

    _____

    _____

    _____

    _____

2. Use a chemical encyclopedia to write the difference between the indicator endpoint for a titration and the equivalence point of the titration.

    _____

    _____

    _____

    _____

    _____

    _____

    _____

    _____

    _____

3. The true equivalence point in these titrations was at pH > 7. What is the pH range for the color change of phenolphthalein? Explain why phenolphthalein could be used for such titrations.

    _____

    _____

    _____

    _____

    _____

    _____

EXPERIMENT 27

Buffered Solutions

Objective

Buffering of weak acid-weak base solutions is very important, especially in biological chemistry. In this experiment you will demonstrate the buffer effect and investigate a situation in which a buffered solution may arise.

Introduction

A buffered solution is a solution that does not change its pH significantly when a strong acid or base is added to it. Buffered solutions typically consist of an approximately equimolar mixture of a conjugate acid-base pair. For example, the following mixtures would be expected to act as buffered solutions:

$0.10\ M\ HC_2H_3O_2$ (acid) and $0.10\ M\ Na^+C_2H_3O_2^-$ (conjugate base)

$0.25\ M\ NH_3$ (base) and $0.20\ M\ NH_4^+Cl^-$ (conjugate acid)

The two components of the buffered solution do not have to be present in exactly equal amounts, but there must be comparable amounts of the components for the buffer to have a significant capacity to resist changes in its pH.

A buffered solution is able to resist changes in its pH when strong acids or strong bases are added because the components of the buffered solution are able to react chemically with such added substances. If the added strong acid or strong base is chemically consumed by one of the components of the buffered solution, then the acid or base will not have an effect on the total hydrogen ion concentration (pH) of the solution.

For example, for the $HC_2H_3O_2/NaC_2H_3O_2$ buffer above, the weak acid portion of the buffer will react with any added strong base, and the conjugate base portion of the buffer will react with any added strong acid:

$HC_2H_3O_2 + NaOH$ (strong base) $\rightarrow NaC_2H_3O_2 + H_2O$

$C_2H_3O_2^- + HCl$ (strong acid) $\rightarrow HC_2H_3O_2 + Cl^-$

Buffered solutions are vitally important in the physiology of living cells. Many biochemical reactions are extremely sensitive to pH, and will not take place if the acidity of the physiological system is outside a very narrow range. For example, if a few drops of acid are added to whole milk, the milk will almost instantly "curdle" as the protein in the milk precipitates. The change in the pH is enough to cause the disruption of the structure of the milk protein so that it is no longer soluble in water. In this experiment you will demonstrate the ability of buffered solutions to resist changes in pH, and will investigate a common situation in which buffered solutions arise.

| Safety Precautions | • Safety eyewear approved by your institution must be worn at all times while you are in the laboratory, whether or not you are working on an experiment. |
| | • Assume that all the acid and base solutions used in this experiment are corrosive to eyes, skin and clothing. Wash immediately if spilled and inform the instructor. Clean up all spills on the benchtop. |

Apparatus/Reagents Required

- pH 7 buffer concentrate
- universal indicator and color chart
- 3 M HCl
- 3 M NaOH
- dropper bottles of 1.0 M HCl, 1.0 M acetic acid, and 1.0 M NaOH, 0.1 M acetic acid, 0.1 M sodium acetate
- albumin (egg white)
- toothpicks

Procedure

1. The Buffer Effect

Place approximately 25 mL of distilled water into each of two small beakers.

To one of the beakers of distilled water, add 1 mL of concentrated pH 7.00 standard reference buffer.

Add two or three drops of universal indicator to each of the beakers. Record the color of the liquid in each beaker after the indicator is added. Use the color chart provided for the indicator to record the pH of the water and of the buffered solution.

To each beaker, add one drop of 3 M HCl solution and stir to mix. Record the color of the solution in each beaker. Use the indicator color chart to record the pH of the two solutions. Which solution underwent the larger change in its pH when the acid was added?

To each beaker, add 2 drops of 3 M NaOH solution (one drop of the NaOH is to neutralize the acid that was added previously) and stir to mix. Record the color of the solution in each beaker. Use the color chart provided for the indicator to record the pH of each solution. Which solution underwent the larger change in its pH when the base was added?

2. Buffering During Titrations

In Part 1 above, you demonstrated the buffer effect on a buffered solution that was purposely prepared. Buffered solutions may also arise *in situ* during titrations of weak acids or bases, which may make the endpoint of the titration less sharp. If we were to titrate a solution of the strong acid HCl with a sodium hydroxide solution, the pH would change drastically and suddenly when we had reached the point at which sufficient NaOH had been added to react with all the HCl present. On the other hand, if we were to titrate a solution of the weak acid acetic acid with sodium hydroxide, the change in pH at the endpoint would be much less sudden and drastic. As sodium hydroxide is added to acetic acid, the acetic acid is converted to sodium acetate, and a buffered solution arises.

Set up two clean, dry test tubes in a rack. Add 20 drops of 1.0 M HCl to one of the test tubes and 20 drops of 1.0 M acetic acid to the other.

Add one drop of universal indicator to each of the test tubes and mix. Use the color chart provided with the indicator to record the pH of each of the solutions before any change is made to them.

Obtain a dropper bottle of 1.0 M NaOH solution, and prepare to titrate each of the acid samples. Use the indicator color chart to record the pH of the solutions after each drop of NaOH has been added (give your best estimate of pH if the color appears intermediate between two of the colors on the chart).

Titrate the acid samples in parallel so that you can compare them: that is, add a drop of NaOH to the HCl solution and record the pH, then add a drop of NaOH to the acetic acid solution and record its pH. Use a stirring rod or toothpick to stir the samples after each drop of NaOH has been added.

Continue adding NaOH dropwise until the pH has reached a value of 10.

Using graph paper from the back of this manual, construct two graphs (one for each titration): plot the pH of the solution at each point in the titration versus the number of drops of NaOH that had been added to reach that point. Draw a smooth curve through your data points (do not simply connect the dots).

Notice the difference in shape between the two curves: the strong acid/strong base titration shows a sharp, sudden increase in pH after approximately 20 drops of the NaOH had been added. The weak acid/strong base titration shows a much more gradual change in pH as the NaOH was added, reflecting the fact that the system was buffering.

3. Proteins as Buffers

Proteins are constructed of long chains of linked amino acid molecules. An amino acid molecule contains both a weak acid group (the carboxyl group) and a weak base group (the amino group). Since amino acids have both acidic and basic properties, solutions of proteins behave as buffered solutions when strong acids or strong bases are added.

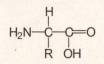

A general amino acid

Place approximately 10 mL of water in a small beaker. Add a tiny amount of egg albumin – approximately the size of the head of a match. Stir the mixture vigorously to dissolve as much albumin as possible. Decant approximately 5 mL of the albumin solution into a clean test tube (avoid transferring any un-dissolved albumin from the beaker to the test tube.)

Set up the test tube containing the albumin and a second test tube containing an equal amount of water in a test tube rack. Add one drop of universal indicator to each of the test tubes and mix. Use the color chart to record the pH of each of the solutions before any change is made to them.

Add one drop of 1 M HCl to each test tube. Record the color of the indicator. Then add two drops of 1 M NaOH to each test tube. Record the color of the indicator. Did the albumin solution appear to behave as a buffered solution?

4. Buffer Capacity

Set up two small test tubes in a test tube rack and label them as 1 and 2.

To test tube 1, add five drops of 0.1 M acetic acid and 5 drops of 0.1 M sodium acetate.

To test tube 2, add one drop of 0.1 M acetic acid, one drop of 0.1 M sodium acetate and eight drops of distilled water.

Stir the solutions with a clean stirring rod or toothpick. Add one drop of universal indicator to each test tube, stir, and record the pH of the solutions. The solutions should be acidic.

Using a stirring rod or toothpick to stir the solution, add 0.1 M NaOH to test tube 1, one drop at a time, counting the number of drops of NaOH solution required to reach pH 7 (or higher).

Repeat, this time with the solution in test tube 2.

Which solution required fewer drops of NaOH to reach pH 7 (or higher)? Why?

EXPERIMENT 27

Buffered Solutions

Pre-Laboratory Questions

1. Use your textbook to write a specific definition for a buffered solution.

   _____

   _____

   _____

   _____

   _____

   _____

   _____

   _____

   _____

2. Give two examples of mixtures that would behave as buffered solutions, and show how (by writing equations) the components of each of your solutions would consume strong acid (HCl) and strong base (NaOH) if added to the solution. Do not use examples discussed in either this lab manual or in your textbook.

   _____

   _____

   _____

   _____

   _____

   _____

   _____

   _____

   _____

   _____

   _____

   _____

3.   Buffered solutions are especially important in biological systems. Why? Give two examples of buffered solutions in biological systems.

_____

_____

_____

_____

_____

_____

_____

_____

_____

_____

4.   Use your textbook, a chemical encyclopedia or the Internet to explain what is meant by the buffer capacity of a buffered solution.

_____

_____

_____

_____

_____

_____

_____

Name: _____ Section: _____

Lab Instructor: _____ Date: _____

EXPERIMENT 27

Buffered Solutions

Results/Observations

1. **The Buffer Effect**

   Color of distilled water + indicator _____

   Color of buffered solution + indicator _____

   pH of distilled water _____ pH of buffered solution _____

   pH after adding HCl: distilled water _____ buffer _____

   pH after adding NaOH: distilled water _____ buffer _____

   Which solution underwent the larger changes in pH when either HCl or NaOH was added? Why?

   _____

   _____

2. **Buffering During Titrations**

   Initial pH of HCl solution _____

Drops	pH	Drops	pH	Drops	pH
1	_____	9	_____	17	_____
2	_____	10	_____	18	_____
3	_____	11	_____	19	_____
4	_____	12	_____	20	_____
5	_____	13	_____	21	_____
6	_____	14	_____	22	_____
7	_____	15	_____	23	_____
8	_____	16	_____	24	_____

Experiment 27:   Buffered Solutions

Initial pH of acetic acid solution _____

Drops	pH	Drops	pH	Drops	pH
1	_____	9	_____	17	_____
2	_____	10	_____	18	_____
3	_____	11	_____	19	_____
4	_____	12	_____	20	_____
5	_____	13	_____	21	_____
6	_____	14	_____	22	_____
7	_____	15	_____	23	_____
8	_____	16	_____	24	_____

3.   Proteins as Buffers

Did the albumin solution appear to exhibit the properties of a buffered solution (compared to water) when strong acid and strong base were added? Explain your observations.

_____

_____

_____

_____

4.   Buffer Capacity

Number of drops NaOH to reach pH 7 for test tube 1   _____

Number of drops NaOH to reach pH 7 for test tube 2   _____

Explanation

_____

_____

_____

_____

Questions

1.   Explain/discuss the "buffer effect" you demonstrated in Part 1 of the experiment.

_____

_____

_____

_____

_____

_____

_____

_____

_____

_____

_____

_____

2.   Describe in your own words how the shapes of your two titration curves differ as determined in Part 2 of the experiment. Why does one curve show a sudden, dramatic increase in pH, whereas the other curve shows a much more gradual increase in pH?

_____

_____

_____

_____

_____

_____

_____

_____

_____

_____

_____

3.   In Part 4 of the procedure, you prepared two acetic acid/sodium acetate buffers. The buffered solution in tube 1 contained five times the amount of acetic acid and sodium acetate as the buffered solution in tube 2. Was this difference reflected in the number of drops of NaOH solution each buffered solution sample required before reaching pH 7? Explain.

_____

_____

_____

_____

_____

_____

_____

_____

_____

_____

_____

4.   Given the general structure of an amino acid:

Which portion of the amino acid molecule might be expected to combine with bases? Which portion of the molecule would combine with acids? Explain.

_____

_____

_____

_____

_____

_____

_____

_____

_____

_____

_____

EXPERIMENT 28

Electrolysis

Objective

Electrolysis is the use of an electrical current to force a chemical reaction that would ordinarily not proceed spontaneously. When an electrical current is passed through a molten or dissolved electrolyte, two chemical changes take place. At the anode, an oxidation half-reaction takes place. At the cathode a reduction half-reaction takes place. Exactly what half-reaction occurs at each electrode is determined by the relative ease of oxidation-reduction of all the species present in the electrolysis cell. In this experiment you will study the electrolysis of water and the electrolysis of a solution of the salt potassium iodide.

Introduction

When an electrical current is passed through water, two chemical processes take place. At the anode, water molecules are oxidized:

$$2H_2O \rightarrow O_2 + 4H^+ + 4e^-$$

Gaseous elemental oxygen is produced at the anode and can be collected and tested. During the electrolysis of water, water molecules are reduced at the cathode:

$$4e^- + 4H_2O \rightarrow 2H_2 + 4OH^-$$

Gaseous elemental hydrogen is produced at the cathode and may be collected and tested.

The two processes above are called *half-reactions*. It is the combination of the two half-reactions, which are taking place at the same time but in different locations, that constitutes the overall reaction in the electrolysis cell. The overall cell reaction that takes place is obtained by adding together the two half-reactions and canceling species common to either side:

$$6H_2O \rightarrow O_2 + 2H_2 + 4H^+ + 4OH^-$$

However, when the hydrogen ions and hydroxide ions migrate toward each other in the cell, they will react with each other:

$$H^+ + OH^- \rightarrow H_2O$$

The result is the production of four water molecules. The final overall equation for what occurs in the cell is simply:

$$2H_2O(l) \rightarrow O_2(g) + 2H_2(g)$$

Note the coefficients in this final overall equation. Twice as many moles of elemental hydrogen gas are produced as elemental oxygen gas. If the gases produced are collected, then according to Avogadro's law, the volume of hydrogen collected should be twice the volume of oxygen collected.

The reactions that take place in an electrolysis cell in which more than one reaction is possible are always those that require the least expenditure of energy. For the electrolysis of water discussed above, because water was the only reagent present in any quantity, water was both oxidized at the anode and reduced at the cathode. In the second part of the experiment, you will electrolyze a solution of the salt potassium iodide, KI.

Two possible oxidation half-reactions must be considered. Depending on which species in the solution is more easily oxidized, one of these half-reactions will represent what actually occurs in the cell:

$$2H_2O \rightarrow O_2 + 4H^+ + 4e^-$$

$$2I^- \rightarrow I_2(s) + 2e^-$$

In the first half-reaction, water is being oxidized. This half-reaction would generate elemental oxygen gas, which can be detected with a glowing splint. In addition, the pH of the solution in the region of the anode would be expected to decrease as hydrogen ion is generated by the electrode process. An indicator might be added to determine if the pH changes in the region of the anode. In the second possible half-reaction, elemental iodine is generated. Elemental iodine is slightly soluble in water, producing a brown solution, which you would notice in the solution surrounding the anode. You might also notice the anode becomes coated with dark crystals of solid iodine as the electrolysis continues.

The reduction half-reaction that takes place at the cathode in this experiment could be one of two processes, depending on which reduction requires a lower expenditure of energy:

$$2H_2O + 2e^- \rightarrow H_2 + 2OH^-$$

$$K^+ + e^- \rightarrow K(s)$$

If the reduction of water is the actual half-reaction, as in the first example, hydrogen gas would be generated at the cathode and could be collected and tested for its flammability. Notice that hydroxide ion is also produced. Hydroxide ion will make the solution basic in the region of the cathode. As before, an indicator might be added to detect this change in pH in the region of the cathode. If the actual reduction, on the other hand, involves potassium ion, the cathode would be expected to increase in size (and mass) as potassium metal is plated out on the surface of the cathode.

By careful observation in this experiment, you should be able to determine which oxidation and which reduction actually take place in the cell.

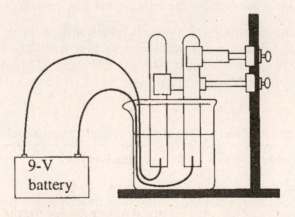

Figure 28-1. Setup for electrolysis of water
The gases produced at the electrodes will be captured in the test tubes for subsequent identification.

Safety Precautions	• Safety eyewear approved by your institution must be worn at all times while you are in the laboratory, whether or not you are working on an experiment.
	• Sodium sulfate and potassium iodide may be irritating to the skin. Wear gloves during their use. Wash after handling.

Apparatus/Reagents Required

- 9-v batteries and "snap-cap" connecting wires
- electrodes (graphite rods or inert metal)
- two test tubes with tightly fitting stoppers for collecting the products
- wood splints
- copper wire with alligator clips
- ruler
- 1 M sodium sulfate solution
- potassium iodide
- sodium thiosulfate
- universal indicator solution and color chart
- starch solution
- steel wool or sandpaper
- disposable gloves

Procedure

Record all data and observations directly in your notebook in ink.

1. Preparation of the Power Source

To avoid any danger of electrical shock, the electrical power needed for the electrolyses will be provided by a 9-v battery. Fit the battery with a "snap cap" connector.

Attach longer copper wires – they will be attached to the electrolysis electrodes – to the wire leads of the "snap-cap" connector on the battery. Connect "alligator clips" to the loose ends of longer wires for use in clipping to the electrodes.

2. Electrolysis of Water

Wear gloves during this procedure.

Place approximately 200 mL of distilled water in a 400-mL beaker. Add about 2 to 3 mL of 1 M sodium sulfate to the water and stir. The sodium sulfate is added to help the electrical current pass more easily through the cell. (The ions in sodium sulfate are much more difficult to oxidize or reduce than are water molecules and do not interfere with the electrolysis of water.)

Examine the electrodes to be used. Most commonly the electrodes will be small graphite rods, or short lengths of an inert metal. If the electrodes are metallic, and appear to be covered with an oxide coating, scrub the surface of the electrode with steel wool or sandpaper and rinse with distilled water.

Arrange the apparatus as indicated in Figure 28-1, *but do not attach the leads from the battery yet*. Make sure that the electrodes do not touch each other, and be certain that the electrodes are arranged in such a way that the test tubes can be inverted over them easily.

Fill each of the test tubes to be used for collecting gas with some of the water to be electrolyzed. Take one of the test tubes and place your finger over the mouth of the test tube to prevent loss of water.

Invert the test tube and lower the test tube into the water in the beaker. Remove your finger, and place the test tube over one of the electrodes so that the gas evolved at the electrode surface will be directed into the test tube.

If the liquid in the test tube is lost during this procedure, remove the test tube, refill with water, and repeat the transfer. Repeat the procedure with the other test tube and the remaining electrode.

Have the instructor check the apparatus before continuing.

If the instructor approves, connect the alligator clips on the battery leads to the electrodes to begin the electrolysis. Allow the electrolysis to continue until one of the test tubes is just filled with gas. Which gas is it?

Disconnect the battery leads from the electrodes.

Stopper the test tubes while they are still under the surface of the water in the beaker and remove. One test tube should be filled with gas (hydrogen), while the other test tube should be only about half-filled with gas (oxygen), with the remainder of the test tube filled with water.

3. Testing of the Gases Evolved

With a ruler, measure the approximate height of gas contained in each test tube as an index of the volume of gas that was generated. Do the relative amounts of hydrogen and oxygen generated seem to correspond to the stoichiometry of the reaction?

Light a burner or match. Using a clamp or test tube holder to protect your hands, hold the test tube containing the hydrogen gas *upside down* (hydrogen is lighter than air) and remove the stopper. Bring the flame near the open mouth of the test tube. Describe what happens to the hydrogen when ignited.

Obtain a wooden splint. Ignite the splint in a burner flame or match; then blow out the splint quickly so that the wood is still glowing. Remove the stopper from the oxygen test tube and insert the splint. Describe what happens to the splint.

4. Electrolysis of Potassium Iodide

Wear gloves during this procedure.

Weigh out approximately 2.5 g of potassium iodide and dissolve in 150 mL of distilled water. You should have an approximately 0.1 M KI solution.

The KI solution should be colorless. If the solution is brown at this point, some of the iodide ion present has been oxidized. If this has happened, add a *single* crystal of sodium thiosulfate and stir. If the brown color does not fade, add more single crystals of sodium thiosulfate until the potassium iodide solution is colorless.

Add four or five drops of universal indicator solution to the beaker and stir. Record the color and pH of the solution. Keep handy the color chart provided with the indicator.

Arrange the electrodes as indicated in Figure 28-1, but *do not connect the battery leads yet*. Make sure that the electrodes do not touch each other and that the electrodes are arranged in such a way that the test tubes can be inverted over them easily.

Fill each of the test tubes with some of the KI solution to be electrolyzed. Take one of the test tubes and place your finger over the mouth of the test tube to prevent loss of solution.

Invert the test tube and lower the test tube into the solution in the beaker. Remove your finger, and place the test tube over one of the electrodes so that the substances evolved at the electrode surface will be directed into the test tube.

If the liquid in the test tube is lost during this procedure, remove the test tube, refill with solution, and repeat the transfer. Repeat the procedure with the other test tube and the remaining electrode.

Have the instructor check the apparatus before continuing.

If the instructor approves, connect the battery leads to the electrodes to begin the electrolysis. Examine the electrodes for evolution of gas or deposition of a solid. Allow the electrolysis to continue for several minutes. (If a gas is generated in the cell reaction, stop the electrolysis when the test tube above the electrode is filled with the gas.)

Observe and record the color changes that take place in the solution in the region of the electrodes. Be careful to distinguish between color changes associated with the indicator and the possible production of elemental iodine (brown color). By referencing the color chart provided with the indicator, determine what pH changes (if any) have occurred near the electrodes.

While still under the surface of the solution in the beaker, stopper the test tubes, and remove them from the solution in the beaker.

The possible oxidation and reduction half-reactions for this system were listed in the introduction. By testing the contents of the two test tubes, determine which half-reactions actually occurred.

If a gas is present in either test tube, use a clamp to hold the test tube and test the gas with a glowing wood splint. If you suspect the gas is hydrogen, invert the test tube (hydrogen is lighter than air), remove the stopper, and bring the wood splint near the mouth of the test tube. Hydrogen will explode with a loud pop. If you suspect the gas is oxygen, hold the test tube upright with the clamp, remove the stopper, and insert the glowing splint. Oxygen will cause the splint to burst into full flame.

If elemental iodine were produced, one of the test tubes would contain a brown solution. Confirm the presence of iodine by addition of a few drops of starch (starch reacts with iodine to produce an intense blue/black color). If iodine was produced, the electrode at the oxidation site would probably be coated with a thin layer of gray-black iodine crystals.

If metallic potassium were produced, it would have plated out as a thin gray-white coating on one of the electrodes.

After determining what oxidation and what reduction have actually occurred in the electrolysis cell, combine the appropriate half-reactions into the overall cell reaction for the electrolysis.

EXPERIMENT 28

Electrolysis

Pre-Laboratory Questions

1. Suppose 25 mL of gaseous hydrogen is collected through the electrolysis of water. What volume of gaseous oxygen should also be collected? Explain why.

_____

_____

_____

_____

_____

_____

_____

_____

_____

2. Describe the qualitative tests for oxygen gas and hydrogen gas you will be performing in this experiment.

_____

_____

_____

_____

_____

_____

_____

_____

_____

_____

_____

3.   Why is a 9-v transistor battery used as the source of electrical power for electrolysis, rather than a
      power-supply that plugs into the wall electrical current?

_____

_____

_____

_____

_____

_____

4.   What do we mean by "half-reactions" when discussing electrochemical processes? What do the
      half-reactions for an overall process represent? When there are several *possible* theoretical half-
      reactions for an electrolysis, which half-reactions will actually occur?

_____

_____

_____

_____

_____

_____

_____

_____

_____

_____

_____

_____

_____

_____

EXPERIMENT 28

Electrolysis

Results/Observations

1. Electrolysis of Water

Observations on electrolysis

_____

_____

_____

Approximately how long did it take to fill the test tube with H_2? _____

Height of gas in O_2 tube _____ in H_2 tube _____

Ratio of H_2/O_2 heights_____ error _____

Observation on testing H_2 with flame _____

Observation on testing O_2 with wood splint _____

2. Electrolysis of Potassium Iodide

Was it necessary to add crystals of sodium thiosulfate to the potassium iodide solution? If so, how many crystals did you add?

_____

Color and pH of KI solution before electrolysis _____

Color and pH of KI solution in region of the anode _____

Color and pH of KI solution in region of the cathode_____

What gas(es) were evolved during the electrolysis? At which electrode? How did the gas(es) respond when tested with the glowing wood splint?

_____

_____

_____

_____

Was elemental iodine produced? How was this finding confirmed?

_____

_____

Was elemental potassium produced? How was its presence confirmed?

_____

_____

Appearance of cathode after electrolysis

_____

_____

Appearance of anode after electrolysis

_____

_____

What half-reactions occurred in the electrolysis of aqueous KI?

_____

_____

What is the overall cell reaction?

_____

Questions

1. Why were you advised to add a crystal of sodium thiosulfate to your potassium iodide solution if it had been brown when prepared? Write a balanced equation for the iodine/thiosulfate reaction.

    _____

    _____

    _____

    _____

    _____

    _____

2. If you had electrolyzed a potassium bromide solution, instead of potassium iodide, what would the likely half-reactions have been? Explain.

    _____

    _____

    _____

    _____

    _____

    _____

    _____

EXPERIMENT 29

Radioactivity

Objective

In this experiment, you will investigate samples that produce the three most common types of nuclear emissions (alpha, beta and gamma) and will study the effect of *shielding* and *distance* on levels of radioactivity.

Introduction

Nuclear processes involve the rearrangement of particles within the nucleus of the atom. If a nucleus is unstable, it may attempt to reach stability by emission of one or more particles and its excess energy. Although there are several particles that may be emitted by unstable nuclei, the most common emissions involve release of *alpha particles* (which are essentially helium nuclei) and *beta particles* (which are high energy electrons produced in the nucleus). Many nuclei also emit excess energy as *gamma* rays (which are similar to x-rays, but of higher energy).

When a nucleus emits an alpha particle (α or ^4_2He), the remaining nucleus will have an atomic number two units lower than the original nucleus, as well as a mass number lower by four atomic mass units. For example, when radium-222 decays, it releases an alpha particle and is itself converted into radon-218:

$$^{222}_{88}Ra \rightarrow \, ^4_2He + \, ^{218}_{86}Rn$$

Notice however, that there is a conservation of both the mass number and the atomic numbers (mass number: $222 = 4 + 218$; atomic number: $88 = 2 + 86$).

When an unstable nucleus emits a beta particle (β or $^0_{-1}e$), the resulting nucleus will have an atomic number one unit higher than the original unstable nucleus, but will have the same mass number. For example, iodine-131 decays by a β emission:

$$^{131}_{53}I \rightarrow \, ^0_{-1}e + \, ^{131}_{54}Xe$$

Notice that in this process also, the mass numbers and atomic numbers are both conserved.

In addition to emitting an alpha or beta particle as shown above, many nuclei also rid themselves of excess energy by emitting gamma radiation. Gamma rays are similar to x-rays in nature (being pure electromagnetic energy), but they tend to be of shorter wavelength (and therefore higher energy) since they arise from transitions within the nucleus. Gamma rays represent photons of electromagnetic radiation, and have zero charge and zero mass number. For example, when uranium-238 undergoes alpha emission, two gamma rays (of different energies) are also emitted with the alpha particles.

Needless to say, nuclear processes can be dangerous. The particles and radiation emitted can cause changes in human cells, which can result in abnormalities forming in current or subsequent generations. In this experiment, you will investigate standard samples of radioisotopes and will detect the various types of radiation they emit with a simple Geiger counter. You will also determine the effect distance has on the level of radioactivity and will study how shielding may be used to stop the various types of radioactive particles.

Experiment 29:   Radioactivity

Safety Precautions	• Safety eyewear approved by your institution must be worn at all times while you are in the laboratory, whether or not you are working on an experiment. • Although the level of radiation emitted by the samples used in this experiment is so low as to not be regulated, you should treat the radioactive samples as if they were highly toxic materials. • Wear gloves during the experiment. Keep your hands away from your mouth and nose at all times. Wash thoroughly after using the radioactive materials. Pregnant women may wish to consult with their physician before performing the experiment

Apparatus/Reagents Required

- Geiger counter and instructions
- standard alpha, beta and gamma sources
- other sources as provided by the instructor (which may include smoke detectors, anti-dust devices, salt substitute, radioactive rocks, radioactive china, etc.)
- aluminum foil
- lead foil
- paper
- yard stick or meter stick

Procedure

Since the number of Geiger counters available is limited, your instructor may ask you to work in small groups for making your determinations. All students in the group should participate in making the measurements and should have a chance to examine the working of the Geiger counter.

Since there are various models of Geiger counter available, your instructor will provide you with specific instructions for the model you will be using. Typically, the orientation of the Geiger counter to the sample is what determines which type of radiation the Geiger counter will be able to detect in that position. Study the instructions carefully and make certain you know how to hold the Geiger counter when making measurements.

Most Geiger counters have both a visual display (either a digital or analog display of the number of counts per minute) as well as an audio feature (clicking or beeping noises when a particle is detected). You will be asked to record the approximate number of counts per minute for each of the samples, and you can use either the visual display or the audio feature (whichever is easiest). The Geiger counter may also have a sensitivity switch: some samples may emit so many particles per minute that it would be impossible to count them directly: the sensitivity switch will enable you to display the number of counts as a power of 10 (for example, if the sensitivity switch is set on "×100", and you detect three counts, then the actual number of particles detected by the Geiger counter was 300).

1. Alpha, Beta and Gamma Sources

Standard radioactive sources that emit primarily alpha particles, beta particles or gamma rays will be provided. These samples contain a small amount of a suitable radioisotope sealed inside a plastic disk. The plastic disk will be labeled with the identity of the radioactive source and the type of radiation it emits.

Examine each radioactive source. Record the identity of the radioactive isotope on the report page.

Use the Geiger counter (see the specific instructions for the Geiger counter provided) to determine the number of counts per minute produced by each source. Hold the Geiger counter as close to the source as possible during this determination. Record.

Write the balanced nuclear equation for the emission for each of the sources studied.

Radioactivity occurs, or is made use of, in several everyday items. Most smoke detectors contain a radioactive isotope as the ionization source. Many glazed bricks used in construction emit low levels of radiation. Devices used to remove dust from vinyl audio recordings may contain a radioisotope. Some pottery or china may be painted or glazed with radioactive materials. If your school has an Earth Science/Geology department, their collection of samples may contain some radioactive ores. Even some commercial salt-substitutes contain radioactivity considerably higher than background levels.

Your instructor may provide you with various additional radioactive sources from everyday life. For these sources, determine the type of emission being produced (alpha, beta or gamma) as well as the number of counts per minute from each source.

2. Effect of Distance on Radiation Level

The simplest means of protecting oneself from radiation is to get as far away from the source of radiation as quickly as possible. The intensity of radiation drops rapidly as the distance from the source increases. In fact, the radiation intensity varies as the inverse of the square of the distance from the source. If you double the distance between you and the radioactive source, the level of radiation is decreased to $1/(2^2) = 1/4$ of its initial value; if you move three times as far from the source, the level of radiation is decreased to about 11% of its initial value ($1/3^2 = 1/9$).

Place a meter or yard stick flat on the bench-top. Choose one of the radioactive sources you have already investigated (preferably one that produced a lot of counts per minute) and place the source as close to the zero end of the meter/yard stick as possible.

Arrange the Geiger counter along the meter/yard stick so that the actual detector portion of the Geiger counter is positioned about 1 cm (~0.5 in.) from the source. Record the exact position. Take a reading of the number of counts per minute at this position. Record.

Move the Geiger counter so that the detector is at 2 cm (~1 in.) from the source. Record the exact position of the Geiger counter and the number of counts per minute in this location.

Move the Geiger counter in additional 1 cm increments along the meter/yard stick, taking readings at each position, until the Geiger counter is 12 cm (~ 6 inches) from the source. Record the exact position of the Geiger counter and the number of counts per minute detected after each change in location.

Does your data seem to confirm the inverse square rule? Explain.

3. Effect of Shielding

The distance radioactive particles can travel through a medium depends both on the nature of the radioactive particles and on the medium through which they are traveling (a dense medium is able to shield better). In particular, we are often concerned with how far radioactive particles can travel through the human body when we are exposed to radiation.

Alpha particles are relatively heavy and large, and cannot travel very far through the body. Most alpha particles are stopped by the skin. Alpha particles are still dangerous however, because they may be inhaled or ingested. Many homeowners test their homes for the presence of Radon gas because of the fear of too-high alpha particle levels in the air. In the laboratory, alpha particles can be stopped by even a single sheet of paper.

Beta particles are less massive than alpha particles, and are expelled from the nucleus with considerable momentum. Their speed enables them to penetrate the outer layers of skin in the body to a depth of perhaps a centimeter. In the laboratory, beta particles would not be stopped by a single sheet of paper. It would require considerably more shielding to stop them.

Gamma radiation consists of pure, high-energy electromagnetic radiation. Similar to x-rays, gamma radiation would pass right through the human body, causing ionization of cell molecules as it passed, making it very dangerous indeed. Gamma radiation is only stopped by a very thick layer of a very dense material. Typically lead is used to shield against gamma radiation. You have undoubtedly noticed that your dentist covers you with a lead apron during dental x-rays to protect your thyroid gland and reproductive organs from the radiation.

Your instructor will have available sheets of paper or cardboard, as well as aluminum and lead foils to be tested as shielding.

Use the standard radioactive sources from Part 1 above to investigate the various types and amount of shielding required to cause the number of counts per minute detected by the Geiger counter to drop to near background levels for each type of radioactive emission.

Place the shielding between the radioactive source and the Geiger counter, and record the identity and thickness of the shielding, as well as the number of counts per minute when the shielding is in place. Experiment with different thicknesses of shielding to determine how much shielding is required to stop each type of radiation from being detected. Record and explain your findings.

Name: _____   Section: _____

Lab Instructor: _____   Date: _____

EXPERIMENT 29

Radioactivity

Pre-Laboratory Questions

1. Write nuclear symbols (in the format $_Z^A X$) for each of the following:

    a.   alpha particle   _____

    b.   beta particle   _____

    c.   neutron   _____

    d.   positron   _____

    e.   the isotope of boron with mass number 10   _____

    f.   the nuclide with mass number 27 and atomic number 13   _____

2. Complete each of the following nuclear equations by supplying the missing particle.

    a. $_4^9 \text{Be} + _1^1 \text{H} \rightarrow ? + _2^4 \text{He}$ _____

    b. $_{88}^{226} \text{Ra} \rightarrow ? + _{86}^{222} \text{Rn}$ _____

    c. $_{92}^{235} \text{U} + ? \rightarrow _{54}^{143} \text{Xe} + _{38}^{90} \text{Sr} + 3\,_0^1 \text{n}$ _____

    d. $_8^{17} \text{O} + ? \rightarrow _6^{14} \text{C} + _1^3 \text{H}$ _____

    e. $_{14}^{28} \text{Si} + _0^1 \text{n} \rightarrow _{12}^{25} \text{Mg} + ?$ _____

3. Although the samples you will deal with in this experiment emit only extremely low levels of
    radioactivity, discuss the precautions you will be taking when dealing with the radioactive sources.

    _____

    _____

    _____

    _____

    _____

    _____

    _____

    _____

    _____

4.   A radioactive sample produces 1,000 counts per minute on a Geiger counter when the sample is placed 1 inch from the Geiger tube. How many counts per minute would you expect if the sample was moved to be 2 inches from the Geiger tube? How many counts per minute at 10 inches from the Geiger tube? Explain.

_____

_____

_____

_____

_____

_____

_____

_____

EXPERIMENT 29

Radioactivity

Results/Observations

1. Alpha, Beta and Gamma Sources

Identity of Source	Type of Emission	Isotope Present	Counts per minute
_____	_____	_____	_____
_____	_____	_____	_____
_____	_____	_____	_____
_____	_____	_____	_____
_____	_____	_____	_____
_____	_____	_____	_____
_____	_____	_____	_____

Equations

_____

_____

_____

_____

_____

2. Effect of Distance on Radiation Level

Distance from source ____ ____ ____ ____ ____

Counts per minute ____ ____ ____ ____ ____

Distance from source ____ ____ ____ ____ ____

Counts per minute ____ ____ ____ ____ ____

Do your data seem to follow the inverse square rule? Explain.

_____

_____

_____

3.   **Effect of Shielding**

Source Used	Particle Emitted	Shielding Used	Thickness Required to Reduce to Background
_____	_____	_____	_____
_____	_____	_____	_____
_____	_____	_____	_____
_____	_____	_____	_____

Explanation of shielding observed

_____

_____

_____

_____

_____

_____

Questions

1.   Although alpha particles are stopped by even a single sheet of paper, they are still very dangerous. Why?

_____

_____

_____

_____

2.   Use your textbook to discuss some of the possible effects of nuclear radiation on the human body.

_____

_____

_____

_____

_____

_____

_____

EXPERIMENT 30

Organic Chemical Compounds

Objective

There are millions of known carbon compounds. In this experiment, several of the major families of carbon compounds will be investigated, which will demonstrate some of the characteristic properties and reactions of each family.

Introduction

Carbon atoms have the correct electronic configuration and size to enable long chains or rings of carbon atoms to be built up. This ability to form long chains is unique among the elements, and is called *catenation*. There are molecules that contain thousands of carbon atoms, all of which are attached by covalent bonds among the carbon atoms. No other element is able to construct such large molecules with its own kind.

Hydrocarbons

The simplest types of carbon compounds are known as *hydrocarbons*. Such molecules contain only carbon atoms and hydrogen atoms. There are various hydrocarbons, differing in the arrangement of the carbon atoms in the molecule, or in the bonding between the carbon atoms. Hydrocarbons consisting of chain-like arrangements of carbon atoms are called *aliphatic hydrocarbons*, whereas molecules containing the carbon atoms in a closed ring shape are called *cyclic hydrocarbons*. The molecules hexane and cyclohexane represent aliphatic and cyclic molecules, each with six carbon atoms:

$$CH_3-CH_2-CH_2-CH_2-CH_2-CH_3 \qquad \begin{array}{c} CH_2-CH_2 \\ / \qquad \backslash \\ H_2C \qquad CH_2 \\ \backslash \qquad / \\ CH_2-CH_2 \end{array}$$

Hexane (left) and cyclohexane (right)

Notice that all the bonds between carbon atoms in hexane and cyclohexane are single bonds. Molecules containing only singly bonded carbon atoms are said to be *saturated*. Other molecules are known in which some of the carbon atoms are attached by double or even triple bonds and are said to be **unsaturated**. The following are representations of the molecules 1-hexene and cyclohexene, both of which contain double bonds:

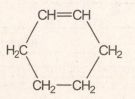

$$CH_2{=}CH-CH_2-CH_2-CH_2-CH_3 \qquad \begin{array}{c} CH{=}CH \\ / \qquad \backslash \\ H_2C \qquad CH_2 \\ \backslash \qquad / \\ CH_2-CH_2 \end{array}$$

1-hexene (left) and cyclohexene (right)

Hydrocarbons with only single bonds are not very reactive. One reaction all hydrocarbons undergo, however, is combustion. Hydrocarbons react with oxygen, releasing heat and/or light. For example, the simplest hydrocarbon is methane, CH_4, which constitutes the major portion of natural gas. Methane burns in air, producing carbon dioxide, water vapor and heat:

$$CH_4 + 2O_2 \rightarrow CO_2 + 2H_2O$$

Single-bonded hydrocarbons will also react slowly with the halogen elements (chlorine and bromine) when exposed to ultraviolet radiation. A halogen atom replaces one (or more) of the hydrogen atoms of the hydrocarbon. For example, if methane is treated with bromine, bromomethane is produced, with concurrent release of hydrogen bromide:

$$CH_4 + Br_2 \rightarrow CH_3Br + HBr$$

The hydrogen bromide produced is generally visible as a fog evolving from the reaction container, produced as the HBr product mixes with water vapor in the atmosphere.

Molecules with double and triple bonds are much more reactive than single-bonded compounds. Unsaturated hydrocarbons undergo reactions at the site of the double (or triple) bond; fragments of some outside reagent attach themselves to the carbon atoms that had been involved in the double (or triple) bond. Such reactions are called *addition reactions*. The orbitals of the carbon atoms that had been used for the extra bond instead are used to attach to the fragments of the new reagent. For example, the double-bonded compound ethylene (ethene), $CH_2{=}CH_2$, will react with many reagents

$$CH_2{=}CH_2 + H_2 \rightarrow CH_3{-}CH_3$$

$$CH_2{=}CH_2 + HBr \rightarrow CH_3{-}CH_2Br$$

$$CH_2{=}CH_2 + H_2O \rightarrow CH_3{-}CH_2OH$$

$$CH_2{=}CH_2 + Br_2 \rightarrow CH_2Br{-}CH_2Br$$

Notice that in all of these reactions of ethylene, the product is saturated and no longer contains a double bond. Many of the reactions of ethylene take place only in the presence of a catalyst, or under reaction conditions that we will not be able to investigate in this experiment.

Cyclic compounds containing only single or double bonds undergo basically the same reactions indicated earlier for aliphatic hydrocarbons. One class of cyclic compounds, called the *aromatic hydrocarbons*, has unique properties that differ markedly from those we have discussed. The parent compound of this group of hydrocarbons is benzene:

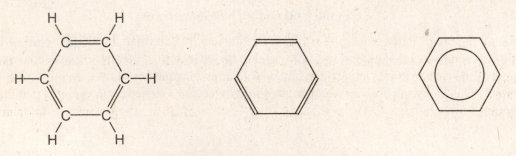

Benzene: three equivalent representations

We typically draw the structure of benzene by indicating double bonds around the ring of carbon atoms. In terms of its reactions, however, benzene behaves as though it has no double bonds. For example, if elemental bromine is added to benzene (with a suitable catalyst), benzene does not undergo an addition reaction as would be expected for a compound with double bonds. Rather, it undergoes a substitution reaction of the sort that would be expected for single-bonded hydrocarbons:

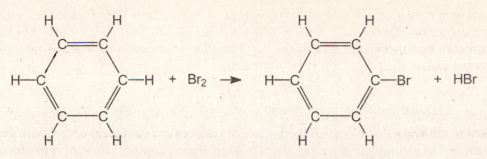

Benzene (left) and bromobenzene (right)

This reaction indicates that perhaps benzene does not have any double bonds in reality. Several Lewis dot structures can be drawn for benzene, with the double bonds in several different locations in the ring. Benzene exhibits the phenomenon of *resonance*; that is, the structure of benzene shown, as well as any other structure that might be drawn on paper, does not really represent the true electronic structure of benzene. The electrons of the double bonds in benzene are, in fact, *delocalized* around the entire ring of the molecule and are not concentrated in any particular bonds between carbon atoms. The ring structure of benzene is part of many organic molecules, including some very common biological compounds.

Functional Groups

While the hydrocarbons are of interest, the chemistry of organic compounds containing atoms other than carbon and hydrogen is far more varied. When another atom, or group of atoms, is added to a hydrocarbon framework, the new atom frequently imparts its own very strong characteristic properties to the resulting molecule. An atom or group of atoms that is able to impart new and characteristic properties to an organic molecule is called a *functional group*. With millions of carbon compounds known, it has been helpful to chemists to divide these compounds into *families*, based on what functional group they contain. Generally, all the members of a family containing a particular functional group will have many similar physical properties and will undergo similar chemical reactions.

Alcohols

Simple hydrocarbon chains or rings that contain a hydroxyl group bonded to a carbon atom are called *alcohols*. The following are alcohols:

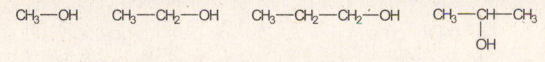

From left to right: methanol, ethanol, 1-propanol, and 2-propanol

The –OH group of alcohols should not be confused with the hydroxide ion, OH⁻. Rather than being basic, the hydroxyl group makes the alcohols, in many ways, act like water, H–OH. For example, the smaller alcohols are fully miscible with water in all proportions. Alcohols are similar to water in that they also react with sodium, releasing hydrogen. Alcohols, however, react more slowly with sodium than water does.

Many alcohols can be oxidized by reagents such as permanganate ion or dichromate ion; the product depends on the structure of the alcohol. For example, ethyl alcohol is oxidized to acetic acid: when wine spoils, vinegar is produced. On the other hand, 2-propanol is oxidized to propanone (acetone) by such oxidizing agents:

Oxidation of primary (ethanol) and secondary (2-propanol) alcohols

Alcohols are very important in biochemistry, since many biological molecules contain the hydroxyl group. For example, carbohydrates (sugars) are alcohols, with a particular carbohydrate molecule generally containing several hydroxyl groups.

Aldehydes and Ketones

An oxygen atom that is double-bonded to a carbon atom of a hydrocarbon chain is referred to as a *carbonyl group*. Two families of organic compounds contain the carbonyl function. The difference between the families has to do with where along the chain of carbon atoms the oxygen atom is attached. Molecules with the oxygen atom attached to the first (terminal) carbon atom of the chain are called *aldehydes*, whereas molecules having the oxygen atom attached to an interior carbon atom are called *ketones*.

Aldehydes are *intermediates* in the oxidation of alcohols. For example, when ethyl alcohol is oxidized, the product is first acetaldehyde, which is then further oxidized to acetic acid:

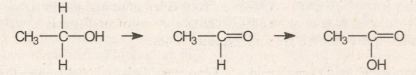

Oxidation of ethanol to acetaldehyde (mild oxidation) and acetic acid (strong oxidation)

Ketones are the end product of the oxidation of alcohols in which the hydroxyl group is attached to an interior carbon atom (called secondary alcohols), as was indicated earlier for 2-propanol.

Aldehydes and ketones are important biologically because all carbohydrates are also either aldehydes or ketones, as well as being alcohols. The common sugars glucose and fructose are shown next. Notice that glucose is an aldehyde, whereas fructose contains the ketone functional group.

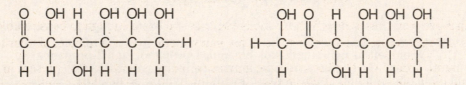

Glucose (left) and fructose (right)

Molecules such as glucose and fructose, which contain both the carbonyl and hydroxyl function, are easily oxidized by Benedict's reagent. Benedict's reagent is a specially prepared solution of copper(II) ion that is reduced when added to certain sugars, producing a precipitate of red copper(I) oxide. Benedict's reagent is the basis for the common test for sugars in the urine of diabetics. A strip of paper or tablet containing Benedict's reagent is added to a urine sample. The presence of sugar is confirmed if the color of the test reagent changes to the color of copper(I) oxide. Benedict's reagent reacts only with aldehydes and ketones that also contain a correctly located hydroxyl group, but will not react with simple aldehydes or ketones.

Organic Acids

Molecules that contain the carboxyl group at the end of a carbon atom chain behave as acids. For example, the hydrogen atom of the carboxyl group of acetic acid ionizes as:

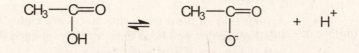

Ionization of acetic acid

On paper, the carboxyl group looks like a cross between the carbonyl group of aldehydes and the hydroxyl group of alcohols, but the properties of the carboxyl group are completely different from either of these. The hydrogen atom of the carboxyl group is lost by organic acids because the remaining ion has more than one possible resonance form, which leads to increased stability for this ion, relative to the non-ionized molecule.

Organic acids are typically weak acids, with only a few of the molecules in a sample being ionized at any given time. However, organic acids are strong enough to react with bicarbonate ion or to cause an acidic response in pH test papers or indicators.

Organic Bases

There are several types of organic compounds that show basic properties, the most notable of these compounds forms the family of *amines*. Amines are considered to be organic derivatives of the weak base ammonia, NH_3, in which one or more of the hydrogen atoms of ammonia is replaced by chains or rings of carbon atoms. The following are some typical amines:

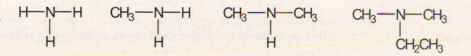

From left to right: ammonia, methylamine, dimethylamine, ethyldimethyamine

Like ammonia, organic amines react with water, releasing hydroxide ion from the water molecules:

$$NH_3 + H_2O \rightleftharpoons NH_4^+ + OH^-$$

$$CH_3-NH_2 + H_2O \rightleftharpoons CH_3NH_3^+ + OH^-$$

Although amines are weak bases (not fully dissociated), they are basic enough to cause a color change in pH test papers and indicators, and are able to complex many metal ions in a manner similar to ammonia.

The so-called amino group, $-NH_2$, found in some amines (primary amines) is also present in the class of biological compounds called *amino acids*. Many of the amino acids from which proteins are constructed are remarkably simple, considering the overall complexity of proteins. The amino acids found in human protein contain a carboxyl group, with an amino group attached to the carbon atom next to the carboxyl group in the molecule's carbon chain. This sort of amino acid is referred to as an alpha amino acid. One simple amino acid found in human protein is shown here:

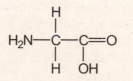

Glycine (2-aminoethanoic acid)

Safety Precautions	• Safety eyewear approved by your institution must be worn at all times while you are in the laboratory, whether or not you are working on an experiment.
	• Assume that all the organic substances are highly flammable. Use only small quantities, and keep them away from open flames. Use a hot water bath if a source of heat is required.
	• Assume that all organic substances are toxic and are capable of being absorbed through the skin. Avoid contact and wash after use. Inform the instructor of any spills.
	• Assume that the vapors of all organic substances are harmful. Use small quantities and work in the fume exhaust hood.
	• Sodium reacts violently with water, liberating hydrogen gas and enough heat to ignite the hydrogen. Do not dispose of sodium in the sinks. Add any excess sodium to a small amount of ethyl alcohol, and allow the sodium to react fully with the alcohol before disposal.
	• Potassium permanganate, potassium dichromate and Benedict's reagent are toxic. Permanganate and dichromate can harm skin and clothing. Exercise caution and wash after use. Chromium compounds are suspected mutagens/carcinogens.
	• Bromine causes extremely bad burns to skin, and its vapors are extremely toxic. Work with bromine only in the fume exhaust hood, and handle the bottle with a towel while dispensing.
	• Some of the chemicals in this experiment may be environmental hazards. Do not pour down the drain. Dispose of all materials as directed by the instructor.

Apparatus/Reagents Required

- hexane
- 1-hexene
- cyclohexane
- cyclohexene
- toluene
- ethyl alcohol
- methyl alcohol
- isopropyl alcohol
- *n*-butyl alcohol
- *n*-pentyl alcohol
- *n*-octyl alcohol
- 10% acetaldehyde solution

- acetone
- 5% glucose solution
- 5% fructose solution
- acetic acid
- propionic acid
- butyric acid
- ammonia
- 10% butylamine
- 1% bromine in methylene chloride (dichloromethane)
- 1% aqueous potassium permanganate
- Benedict's reagent
- sodium pellets (instructor only)
- 5% acidified potassium dichromate solution
- pH paper
- 10% sodium bicarbonate solution
- 1 *M* copper sulfate solution
- gloves

Procedure

Record all data and observations directly on the report pages in ink.

1. Hydrocarbons

Your instructor will ignite small portions of hexane, 1-hexene, cyclohexane, cyclohexene, and toluene to demonstrate that they burn readily (do *not* attempt this yourself). Notice that toluene (an aromatic compound) burns with a much sootier flame than the other hydrocarbons, a property that is characteristic of aromatics.

Obtain small (5-drop) portions of hexane, 1-hexene, cyclohexane, cyclohexene and toluene in separate small test tubes. Take the test tubes to the exhaust hood, and add a few drops of 1% bromine solution (*Caution!*) to each. Those samples containing double bonds will decolorize the bromine immediately (addition reaction).

Those samples that did not decolorize the bromine immediately will gradually lose the bromine color (substitution reaction).

Expose those test tubes that did not immediately decolorize the bromine to bright sunlight, an ultraviolet lamp, or a high-intensity incandescent lamp. The reaction is sensitive to wavelengths of ultraviolet light. Blow across the open mouth of the test tubes to see if a fog of hydrogen bromide becomes visible.

Obtain additional 5-drop samples of hexane, 1-hexene, cyclohexane, cyclohexene, and toluene in separate small test tubes. Add a few drops of 1% potassium permanganate to each test tube, stopper, and shake for a few minutes. Those samples containing double bonds will cause a change in the purple color of permanganate.

2. Alcohols

Obtain 5-drop samples of methyl, ethyl, isopropyl, *n*-butyl, *n*-pentyl and *n*-octyl alcohols in separate clean test tubes. Add an equal quantity of distilled water to each test tube, stopper, and shake.

Record the solubility of the alcohols in water. In alcohols with only a few carbon atoms, the hydroxyl group contributes a major portion of the molecule's physical properties, making such alcohols miscible with water. As the carbon-atom chain of an alcohol gets larger, the alcohol becomes more like a hydrocarbon in nature and less soluble in water.

Obtain 5-drop samples of methyl, ethyl, isopropyl, *n*-butyl, *n*-pentyl and *n*-octyl alcohols in separate clean test tubes.

Wearing disposable gloves, add 10 drops of acidified potassium dichromate to each sample, transfer to a hot water bath, and heat until a reaction is evident. Dichromate ion oxidizes primary and secondary alcohols.

Place approximately 2 mL of isopropyl alcohol in a medium-sized dry test tube.

Bring the test tube containing the isopropyl alcohol to your instructor, and have the instructor add a tiny pellet of sodium to the test tube.

Wait until the sodium in the test tube has *completely reacted*. Then add about 2 mL of distilled water to the test tube, and test the resulting solution with pH paper. The alkoxide ion that remains from the alcohol after reaction with sodium is strongly basic in aqueous solution.

3. Aldehydes and Ketones

Obtain 5-drop samples of acetaldehyde, acetone, glucose and fructose in separate test tubes.

Wearing gloves, add 10 drops of acidified potassium dichromate solution to each sample, and transfer the test tubes to a hot water bath. Aldehydes are oxidized by dichromate, being converted into organic acids.

Record which samples change color.

Obtain 5-drop samples of acetaldehyde, acetone, glucose and fructose in separate test tubes.

To each test tube add 2 mL of Benedict's reagent, and transfer to a hot water bath. Molecules containing a hydroxyl group next to a carbonyl group on a carbon atom chain will cause a color change in Benedict's reagent. This test is applied to urine for the detection of sugar.

4. Organic Acids

Obtain 5-drop samples of acetic acid, propionic acid and butyric acid in separate small test tubes.

Add 2 mL of 10% sodium bicarbonate solution to each test tube: a marked fizzing will result as the acids release carbon dioxide from the bicarbonate.

Obtain 5-drop samples of acetic acid, propionic acid and butyric acid in separate small test tubes.

Add 2 mL of water to each test tube, and test the solution with pH test paper.

5. Organic Bases

Obtain small samples of aqueous ammonia and of *n*-butylamine solution in separate small test tubes.

Determine the pH of each sample with pH test paper.

Add a few drops of 1 *M* copper sulfate solution to each test tube and shake. The color of the copper sulfate solution should darken as the metal/amine complex is formed.

EXPERIMENT 30

Organic Chemical Compounds

Pre-Laboratory Questions

Draw structural formulas for each of the following substances. If the common name of the substance is given, write the IUPAC name for the substance.

1.　hexane   IUPAC name _____

2.　isopropyl alcohol   IUPAC name _____

3.　l-hexene   IUPAC name _____

4.　formaldehyde   IUPAC name _____

5.　cyclohexane   IUPAC name _____

6.　acetaldehyde   IUPAC name _____

7. cyclohexene   IUPAC name _____

8. acetone   IUPAC name _____

9. toluene   IUPAC name _____

10. acetic acid   IUPAC name _____

11. methyl alcohol   IUPAC name _____

12. butyric acid   IUPAC name _____

13. ethyl alcohol   IUPAC name _____

14. *n*-butylamine   IUPAC name _____

15. propionic acid   IUPAC name _____

EXPERIMENT 30

Organic Chemical Compounds

Results/Observations

1. Hydrocarbons

    Observation on igniting

    _____

    Which compounds decolorized bromine immediately?

    _____

    Which compounds decolorized bromine on exposure to light?

    _____

    Which compounds produced an HBr fog?

    _____

    Which compounds caused a color change in $KMnO_4$?

    _____

2. Alcohols

    Which alcohols were fully soluble in water?

    _____

    Which alcohols caused a color change when tested with dichromate ion?

    _____

    Observation of reaction of sodium with isopropyl alcohol

    _____

    Approximate pH of water solution of isopropyl alcohol/sodium product

    _____

3. Aldehydes and Ketones

    Which aldehydes/ketones caused a color change when tested with dichromate ion?

    _____

    Which aldehydes/ketones caused a color change when tested with Benedict's reagent?

    _____

4.   Organic Acids

Observation on test with bicarbonate

_____

Approximate pH of aqueous solution of organic acid samples

_____

5.   Organic Bases

Approximate pH of aqueous solutions of ammonia/*n*-butylamine

_____

Observation on adding $CuSO_4$

_____

Questions

1.   When sodium bicarbonate was added to an organic acid sample, carbon dioxide was evolved. Is this test specific for organic acids, or does it apply to acids in general? Write the balanced equation for the process.

_____

_____

_____

2.   When potassium permanganate was shaken with alkenes, the color of the permanganate changed from the purple color of MnO_4 to the brown/black color of manganese(IV) oxide. What sort of reaction must this be? To what is the original alkene converted in the reaction?

_____

_____

_____

3.   When ethyl alcohol was treated with potassium dichromate, the alcohol was oxidized to acetic acid, yet it was indicated that acetaldehyde is an intermediate in this oxidation. How might aldehydes be isolated during such an oxidation, preventing them from being further oxidized to the acid?

_____

_____

_____

_____

_____

EXPERIMENT 31

Ester Derivatives of Salicylic Acid

Objective

Esters are an important class of organic chemical compounds. In this experiment, two esters of salicylic acid with important medicinal properties will be prepared.

Introduction

Two common esters, acetylsalicylic acid and methyl salicylate, are very important over-the-counter drugs. Since ancient times, it has been known that the barks of certain trees, when chewed or brewed as a tea, had *analgesic* (pain-killing) and *antipyretic* (fever-reducing) properties. The active ingredient in such barks was determined to be salicylic acid. When pure salicylic acid was isolated by chemists, however, it proved to be much too harsh on the linings of the mouth, esophagus and stomach for direct use as a drug in the pure state. Salicylic acid contains the phenolic (–OH) functional group in addition to the carboxyl (acid) group, and it is the combination of these two groups that leads to the harshness of salicylic acid on the digestive tract.

Because salicylic acid contains both the organic acid group (–COOH) as well as the phenolic group (–OH), salicylic acid can undergo two separate esterification reactions, depending on whether it is behaving as an acid (through the –COOH group) or as an alcohol analog (through the –OH group). Research was conducted to modify the salicylic acid molecule in such a manner that its desirable analgesic and antipyretic properties would be preserved, but its harshness to the digestive system would be decreased. The Bayer Company of Germany, in the late 1800s, patented an ester of salicylic acid that had been produced by reaction of salicylic acid with acetic acid.

The ester, commonly called acetylsalicylic acid, or by its original trade name (aspirin), no longer has the phenolic functional group. The salicylic acid acts as an alcohol when reacted with acetic acid. Acetylsalicylic acid is much less harsh to the digestive system. When acetylsalicylic acid reaches the intestinal tract, however, the basic environment of the small intestine causes hydrolysis of the ester (the reverse of the esterification reaction). Acetylsalicylic acid is converted back into salicylic acid in the small intestine and is then absorbed into the bloodstream in that form. Aspirin tablets sold commercially generally contain binders (such as starch) that help keep the tablets dry and prevent the acetylsalicylic acid in the tablets from decomposing into salicylic acid. Since the other component in the production of acetylsalicylic acid is acetic acid, one indication that aspirin tablets have decomposed is an odor of vinegar from the acetic acid released by the hydrolysis.

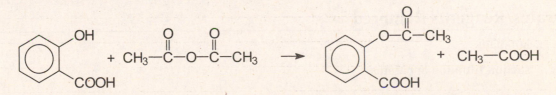

Salicylic acid and acetic anhydride react to produce aspirin and acetic acid.

The second common ester of salicylic acid that is used as a drug is methyl salicylate. When salicylic acid is heated with methyl alcohol, the carboxyl group of salicylic acid is esterified, producing a strong-smelling liquid ester (methyl salicylate).

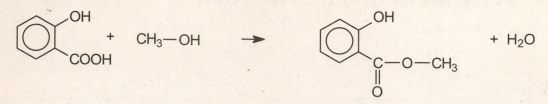

Salicylic acid and methyl alcohol react to produce methyl salicylate and water

The mint odor of many common liniments sold for sore muscles and joints is due to this ester. Methyl salicylate is absorbed through the skin when applied topically and may permit the pain-killing properties of salicylic acid to be localized on the irritated area. Methyl salicylate is a skin irritant, however, and causes a sensation of warmth where it is applied – a desirable property of the ester. Methyl salicylate is also used as a flavoring/aroma agent in various products and is referred to commercially as oil of wintergreen.

Safety Precautions	• Safety eyewear approved by your institution must be worn at all times while you are in the laboratory, whether or not you are working on an experiment.
	• Most of the organic compounds used or produced in this experiment are highly flammable. All heating will be done using a hot plate, and no flames will be permitted in the laboratory.
	• Sulfuric acid is used as the catalyst for the esterification reactions. Sulfuric acid is dangerous and can burn skin, eyes and clothing very badly. If it is spilled, wash immediately before the acid has a chance to cause a burn, and inform the instructor.
	• Acetic anhydride is used in the synthesis of aspirin. Acetic anhydride can seriously burn the skin, and its vapors are harmful to the respiratory tract. If spilled, wash immediately and inform the instructor. Confine the pouring and use of acetic anhydride to the fume exhaust hood.

Apparatus/Reagents Required

- hot plate
- suction filtration apparatus
- ice
- melting point apparatus
- salicylic acid

- acetic anhydride
- methyl alcohol (methanol)
- 50% sulfuric acid
- sodium bicarbonate
- 1 *M* iron(III) chloride

Procedure

Record all data and observations directly on the report pages in ink.

1. Preparation of Aspirin

Set a 250-mL beaker about half full of water to warm on a hot plate in the exhaust hood. A water bath at approximately 70°C is desired, so use the lowest setting of the hotplate control. Check the temperature of the water before continuing.

Weigh out approximately 1 g of salicylic acid and transfer to a clean, dry 125-mL Erlenmeyer flask.

In the exhaust hood, add approximately 3 mL of acetic anhydride (*Caution!*) and two or three drops of 50% sulfuric acid (*Caution!*) to the salicylic acid. Stir until the mixture is homogeneous.

Transfer the Erlenmeyer flask to the beaker of 70°C water (see Figure 31-1) and heat for 15 to 20 minutes, stirring occasionally. Monitor the temperature of the water bath during this time, and do not let the temperature rise above 70°C.

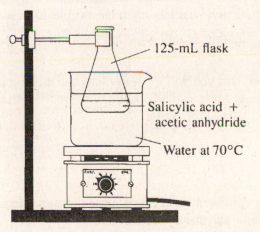

125-mL flask

Salicylic acid + acetic anhydride

Water at 70°C

Figure 31-1. Heating bath for preparation of aspirin and methyl salicylate
A hot plate is used because of the volatility and flammability of several of the reactants and products.

At the end of the heating period, cool the Erlenmeyer flask in an ice bath until crystals begin to form. If crystallization does not take place, scratch the walls and bottom of the flask with a stirring rod to promote formation of crystals.

To destroy any excess acetic anhydride that may be present, add 50 mL of cold water, stir, and allow the mixture to stand for at least 15 minutes to permit the hydrolysis to take place.

Remove the liquid from the crystals by filtering under suction in a Büchner funnel (see Figure 31-2). Wash the crystals with two 10-mL portions of cold water to remove any excess reagents. Continue suction through the crystals for several minutes to help dry them.

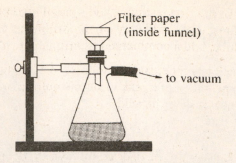

Figure 31-2. Suction filtration apparatus
The filter paper must fit the funnel exactly and must lie flat on the base of the funnel.

You will perform some tests on this crude product today, and your instructor may ask you to save a small portion of your crystals for a melting point determination to be performed during the next laboratory period. If your instructor directs, set aside a small portion of your crystals in a beaker to dry until the next lab period.

Dissolve the remaining portion of the aspirin product in 2 to 3 mL of 95% ethyl alcohol in a 50-mL Erlenmeyer flask. Stir and warm the mixture in the 70°C water bath on the hot plate until the crystals have dissolved completely.

Add approximately 5 mL of distilled water to the Erlenmeyer flask, and heat to re-dissolve any crystals that may have formed.

Remove the Erlenmeyer flask from the hot water bath, and allow it to cool slowly to room temperature. By allowing the solution to cool slowly, relatively large, needle-like crystals of aspirin should form.

Filter the purified product by suction, allowing suction to continue for several minutes to dry the crystals as much as possible.

If your instructor has indicated that you will be performing the melting point determination for your aspirin in the next lab period, set aside a small portion of the recrystallized product in a labeled test tube or beaker to dry until the next laboratory period.

2. Tests on Aspirin

a. Test with Bicarbonate

Acetylsalicylic acid molecules still contain the organic acid group (carboxyl) and will react with sodium bicarbonate to release carbon dioxide gas:

$$H^+ + HCO_3^- \rightarrow H_2O + CO_2$$

Add a very small portion of your aspirin (crude or purified) to a test tube. Add also a small portion of sodium bicarbonate. Add a small amount of water and note the evolution of carbon dioxide. This test indicates only that aspirin is an acid; it is not a specific test for aspirin.

b. Test with Iron(III)

If the synthesis of aspirin has not been effective, or if aspirin has decomposed with time, then free salicylic acid will be present and the product would be harmful if ingested. The standard United States Pharmacopoeia test for the presence of salicylic acid is to treat the sample in question with a solution of iron(III). If salicylic acid is present, the phenolic functional group (–OH) of salicylic acid will produce a purple color with iron(III) ions. The intensity of the purple color is directly proportional to the amount of salicylic acid present.

Set up four test tubes in a rack. To the first test tube, add a small quantity of pure salicylic acid as a control. To the second and third test tubes, add very small portions of your crude and purified aspirin, respectively. To the fourth test tube, add a commercial aspirin tablet (if available).

Add 5 mL of distilled water to each test tube, and stir to mix. Add eight to 10 drops of iron(III) chloride (ferric chloride) solution. The appearance of a pink or purple color in your aspirin or the commercial tablet sample indicates the presence of salicylic acid.

c. Melting Point Determination

After the crude and purified aspirin samples have dried for a week, determine their melting points by the capillary method described in Experiment 4. Compare the observed melting points with the literature value.

3. Preparation of Methyl Salicylate

Place about 1 g (roughly measured) of salicylic acid in a small Erlenmeyer flask and add 5 mL of methyl alcohol. Stir thoroughly to dissolve the salicylic acid.

Add three or four drops of 50% sulfuric acid (*Caution!*) and heat the flask in the 70°C water bath on the hot plate in the exhaust hood for approximately 10 minutes.

Pour the contents of the test tube into approximately 50 mL of warm water in a beaker, and cautiously waft some of the vapors toward your nose. There should be a pronounced odor of wintergreen.

Name: _____ Section:_____

Lab Instructor: _____ Date:_____

EXPERIMENT 31

Ester Derivatives of Salicylic Acid

Pre-Laboratory Questions

1. From the formulas of salicylic acid and acetylsalicylic acid given in the introduction to this experiment, determine their molar masses.

    salicylic acid _____ acetylsalicylic acid _____

2. Suppose 10.0 g of salicylic acid is heated with an excess amount of acetic acid. What is the expected yield of acetylsalicylic acid (aspirin)? Show your calculation.

3. Suppose for the experiment outlined in Question 2, only 6.8 g of aspirin is isolated from the reaction mixture. Calculate the percent yield for the process. Show your work.

4.   Describe the three tests you will be performing on your aspirin product. Discuss what each test is intended to demonstrate about your product.

_____

_____

_____

_____

_____

_____

_____

_____

_____

5.   Why is a hot plate, rather than a gas burner, used for the preparation of aspirin and methyl salicylate?

_____

_____

_____

_____

_____

_____

EXPERIMENT 31

Ester Derivatives of Salicylic Acid

Results/Observations

Observation on crude aspirin (appearance)

_____

_____

Observation of recrystallized aspirin (appearance)

_____

_____

Tests on aspirin

    Effect of bicarbonate _____

    Effect of iron(III) on crude aspirin _____

    Effect of iron(III) on purified aspirin _____

    Effect of iron(III) on salicylic acid _____

    Effect of iron(III) on aspirin tablet _____

Melting points

    Crude aspirin _____ Purified aspirin _____

    Literature value _____ Reference _____

Observations on methyl salicylate _____

Odor? _____

Questions

1.   What acid was used to catalyze the preparation of esters? Could some other acid be used?

    _____

    _____

    _____

    _____

    _____

    _____

    _____

2.   Aspirin was for many years the major household pain-killer, but more recently, other pain-killers such as acetaminophen and ibuprofen have become popular. Find the structures of these substances in a chemical dictionary or handbook. Are these substances esters?

_____

_____

_____

_____

_____

_____

_____

_____

3.   Arrange the various materials you tested with iron(III) in order of increasing purity. How did your impure and recrystallized aspirin samples compare in purity with the commercial aspirin tablet?

_____

_____

_____

_____

_____

_____

_____

_____

_____

_____

4.   Suppose instead of reacting salicylic acid with methyl alcohol (Part 3), you had reacted it with ethyl alcohol, CH_3CH_2OH. Write the equation for the reaction indicating clearly the structure of the ester produced (ethyl salicylate).

_____

_____

_____

_____

_____

_____

_____

EXPERIMENT 32

The Preparation of Fragrant Esters

Objective

Esters are the product of reaction of an organic (carboxylic acid) with an alcohol. Many esters are components of the essential oils of flowers and fruits. Several esters with pleasant fragrances will be synthesized in this experiment, and a common fragrant ester will be hydrolyzed to demonstrate the reverse of the esterification reaction.

Introduction

When an organic acid, R–COOH, is heated with an alcohol, R′–OH, in the presence of a strong mineral acid, the chief organic product is a member of the family of organic compounds known as *esters*.

The general reaction for the esterification of an organic acid with an alcohol is:

$$R–COOH + HO–R' \rightleftharpoons R–CO–OR' + H_2O$$

In this general reaction, R and R′ represent hydrocarbon chains, which may be the same or different. As a specific example, suppose acetic acid, CH_3COOH, is heated with ethyl alcohol, CH_3CH_2OH, in the presence of a mineral acid catalyst. The esterification reaction will be:

$$CH_3–COOH + HO–CH_2CH_3 \rightarrow CH_3–COO–CH_2CH_3 + H_2O$$

The ester product of this reaction ($CH_3–COO–CH_2CH_3$) is named ethyl acetate, indicating the acid and alcohol from which it is prepared. Esterification is an equilibrium reaction, which means that the reaction does not go to completion on its own. Frequently, however, the esters produced are extremely volatile and can be removed from the system by distillation. If the ester is not very easily distilled, it may be possible instead to add a desiccant to the equilibrium system, thereby removing water from the system and forcing the equilibrium to the right.

Unlike many organic chemical compounds, esters often have very pleasant, fruity odors. Many of the odors and flavorings of fruits and flowers are due to the presence of esters in the essential oils of these materials. The table below lists some esters with pleasant fragrances, and indicates from what alcohol and which acid the ester may be prepared.

The esterification reactions shown above are actually equilibrium processes and can be reversed. The reverse of the esterification reaction is referred to as a *hydrolysis* reaction, because it represents the break down of the organic compound through the action of water.

$$R–CO–OR' + H_2O \rightarrow R–COOH + HO–R'$$

Generally a fruit or flower may contain only a tiny amount of ester, giving it a very subtle odor. Usually, the ester is part of some complex mixture of substances, which, taken as a whole, have the aroma attributed to the material. When prepared in the laboratory in relatively large amounts, the ester may seem to have a pronounced "chemical" odor, and it may be difficult to recognize the fruit or flower that has the more nuanced presentation of the aroma.

Table of Common Fragrant Esters

Ester	Aroma	Constituents
n-propyl acetate	pears	*n*-propyl alcohol/acetic acid
methyl butyrate	apples	methyl alcohol/butyric acid
isobutyl propionate	rum	isobutyl alcohol/propionic acid
octyl acetate	oranges	*n*-octyl alcohol/acetic acid
methyl anthranilate	grapes	methyl alcohol/2-aminobenzoic acid
isoamyl acetate	bananas	isoamyl alcohol/acetic acid
ethyl butyrate	pineapples	ethyl alcohol/butyric acid
benzyl acetate	peaches	benzyl alcohol/acetic acid
methyl salicylate	wintergreen	methyl alcohol/salicylic acid

Safety Precautions	• **Safety eyewear approved by your institution must be worn at all times while you are in the laboratory, whether or not you are working on an experiment.** • **Most of the organic compounds used or produced in this experiment are highly flammable. All heating will be done using a hot plate, and no flames will be permitted in the laboratory.** • **Sulfuric acid is used as the catalyst for the esterification reactions. Sulfuric acid is dangerous and can burn skin, eyes and clothing very badly. If it is spilled, wash immediately before the acid has a chance to cause a burn, and inform the instructor.** • **The vapors of the esters produced in this experiment may be harmful. When determining the odors of the esters produced in the experiment, do not deeply inhale the vapors. Merely waft a small amount of vapor from the ester toward your nose.** • **Dispose of all materials as directed by the instructor.**

Apparatus/Reagents Required

- hot plate
- 50% sulfuric acid
- assorted alcohols and organic acids, as provided by the instructor, for the preparation of fruit and flower aromas
- short-stem disposable plastic pipets
- 20% NaOH solution
- methyl salicylate

Procedure

1. Preparation of Fragrant Esters

Set up a water bath in a 250-mL beaker on a hot plate in the exhaust hood. Most of the reactants and products in this choice are highly flammable, and no flames are permitted in the lab during this experiment. Adjust the heating control to maintain a temperature of around 70°C in the water bath.

Some common esters, and the acids/alcohols from which they are synthesized, were indicated in the table in the Introduction. Synthesize at least two of the esters, and note their aromas. Different students might synthesize different esters, as directed by the instructor, and compare the odors of the products.

To synthesize the esters, mix three or four drops (or approximately 0.1 g if the acid is a solid) of the appropriate acid with three or four drops of the indicated alcohol on a clean, dry watch glass.

Add one drop of 50% sulfuric acid to the mixture on the watch glass (*Caution!*).

Use the tip of a plastic pipet to stir the mixture on the watch glass, and then suck as much as possible of the mixture into the pipet.

Place the pipet, tip upward, into the warm-water bath, and allow it to heat for approximately five minutes.

After the five-minute heating period, squirt the resulting ester from the pipet into a beaker of warm water, and cautiously waft the vapors toward your nose.

Remember that the odor of an ester is very concentrated. Several sniffs may be necessary for you to identify the odor of the ester. Record which esters you prepared and their aromas.

2. Hydrolysis of an Ester

Esters may be destroyed by reversing the esterification reaction: water and an ester will react with one another (hydrolysis) to give an alcohol and an organic acid. This reaction is often carried out using sodium hydroxide as catalyst (resulting in the sodium salt of the organic acid, rather than the acid itself).

Set up a water bath in a 250-mL beaker on the hot plate in the fume hood. Adjust the heating control to maintain a temperature of approximately 70°C in the water bath.

Obtain 10 drops of methyl salicylate in a clean test tube. Note the odor of the ester by cautiously wafting vapors from the test tube toward your nose.

Add 2 to 3 mL of water to the test tube, followed by 10 to 15 drops of 20% NaOH solution (*Caution!*).

Heat the methyl salicylate/NaOH mixture in the water bath for 15 to 20 minutes. After this heating period, again note the odor of the mixture. If the odor of mint is still noticeable, heat the test tube for an additional 10-minute period.

Write an equation for the hydrolysis of methyl salicylate.

EXPERIMENT 32

The Preparation of Fragrant Esters

Pre-Laboratory Questions

1. Using R–COOH to represent an organic acid and R′–OH to represent an alcohol, write a general equation for the formation of an ester. What is the other product of the esterification reaction?

_____

_____

_____

2. Draw the structural formula for each of the following esters.

   a.   *n*-propyl acetate

   b.   methyl butyrate

   c.   isobutyl propionate

   d.   octyl acetate

   e.   isoamyl acetate

3.   Why are no flames permitted in the laboratory during this experiment?

_____

_____

_____

_____

_____

_____

_____

4.   Write an equation describing the hydrolysis of methyl salicylate in aqueous sodium hydroxide solution.

_____

_____

_____

_____

EXPERIMENT 32

The Preparation of Fragrant Esters

Results/Observations

1. Preparation of Fragrant Esters

   Which esters did you prepare?

   Ester 1 _____   Ester 2 _____

   Odor _____   Odor _____

   Formula_____   Formula _____

   How do the odors of the esters you produced compare to the "natural" aromas you may have been expecting?

   _____

   _____

   _____

2. Hydrolysis of an Ester

   Odor of methyl salicylate before hydrolysis _____

   Odor after hydrolysis _____

   Equation for hydrolysis reaction _____

Questions

1. Ordinarily esterification reactions come to equilibrium before the full theoretical yield of ester is realized. Aside from distillation discussed earlier, how might one experimentally shift the equilibrium of the esterification reaction so that a larger amount of ester might be isolated?

   _____

   _____

   _____

   _____

   _____

   _____

   _____

   _____

2.   How do chemists define hydrolysis?

_____

_____

_____

_____

_____

_____

_____

_____

3.   Most esters are very volatile and boil at relatively low temperatures. What about the structure of
     ester molecules might account for their relatively low boiling points and high volatility? Hint: Why
     does water have a much higher boiling point than would be expected for such a small molar mass?

_____

_____

_____

_____

_____

_____

_____

_____

_____

_____

_____

_____

Proteins

Objective

In this experiment you will study some of the tests that are used to identify and characterize proteinaceous materials. You will also investigate how a change in environmental conditions can affect a protein.

Introduction

Proteins are long-chain polymers of the α-amino acids:

$$H_2N-\underset{\underset{R}{|}}{\overset{\overset{H}{|}}{C}}-\overset{\overset{}{\underset{OH}{}}}{C}=O$$

They make up about 15% of our bodies. Proteins have many functions in the body. Some proteins form the major structural feature of muscles, hair, fingernails, and cartilage. Other proteins help to transport molecules through the body, fight infections or act as catalysts (enzymes) for biochemical reactions in the cells of the body. Proteins have several levels of structure; each level is very important to a protein's function in the body.

The primary structure of a protein is the particular sequence of amino acids in the polymer chain: each particular protein has a unique primary structure. The secondary structure of a protein describes the basic arrangement in space of the overall chain of amino acids: some protein chains coil into a helical form, while other proteins chains may bind together side-by-side to form a sheet of protein. The tertiary structure of a protein describes how the protein chains (whether sheet or helix) fold in space through interactions of the side groups of the amino acids with each other and with the surroundings: for example, the helical secondary structures of some proteins fold into a globular (spherical) shape to enable them to travel more easily through the bloodstream.

The various levels of structure of a protein are absolutely crucial to the protein's function in the body. For example, if any error occurs in the primary structure of the protein, a genetic disease may result (sickle cell disease and Tay-Sachs disease are both due to errors in protein primary structures). If any change in the environment of a protein occurs, the tertiary and secondary structures of the protein may be changed enough to make the protein lose its ability to act in the body. For example, egg white is basically an aqueous solution of the protein albumin: when an egg is cooked, the increased temperature causes irreversible changes in the tertiary and secondary structures of the albumin, making it coagulate as a solid.

When the physical or chemical environment of a protein is modified, and the protein responds by changing its tertiary or secondary structure, the protein is said to have been *denatured*. Denaturing may be reversible (if the change to the protein's environment is not too drastic) or irreversible (as with the egg white discussed above). Changes in environment that may affect a protein's structure include: changes in temperature, changes in the pH (acidity) of the protein's environment, changes in the solvent in which the protein is suspended, or the presence of any other ions or molecules that may chemically interact with the protein (such as ions of the heavy metals) and cause the protein to lose its function.

	•	Safety eyewear approved by your institution must be worn at all times while you are in the laboratory, whether or not you are working on an experiment.
Safety Precautions	•	Sodium hydroxide, nitric acid and hydrochloric acid solutions are corrosive to skin, eyes, and clothing. Wash immediately if spilled, and clean up all spills on the bench top.
	•	Lead acetate and copper(II) sulfate solutions are toxic if ingested. Wash after handling.
	•	Some chemicals in this solution may be environmental hazards. Do not pour down the drain. Dispose of all materials as directed by the instructor.

Apparatus/Reagents Required

- egg white (1% albumin) solution
- nonfat milk
- plain gelatin, cut into small cubes
- 1% alanine
- 10% NaOH
- 3% copper(II) sulfate
- 6 M HNO$_3$
- 6 M NaOH
- 5% lead(II) acetate
- 3 M HCl
- saturated NaCl solution
- ethyl alcohol

Procedure

1. Tests for Proteins

Samples of three proteinaceous materials will be available for testing: egg white (containing the protein albumin); nonfat milk (containing the protein casein) and gelatin. Samples of the free amino acid alanine will also be tested for comparison and contrast.

a. Biuret test

The biuret test detects the presence of proteins and polypeptides, but does not detect free amino acids (the test is specific for the peptide *linkage*).

Place about 10 drops of 1% albumin, nonfat milk and 1% alanine solutions, along with a small cube of gelatin in separate clean, small, test tubes.

Add 10 drops of 10% sodium hydroxide solution *(Caution!)* to each test tube, followed by three or four drops of 3% copper(II) sulfate solution.

Carefully mix the contents of each test tube with a clean stirring rod, being sure to clean the stirring rod when switching between solutions. Make sure to break up the gelatin sample as much as possible when mixing.

The blue color of the copper(II) sulfate solution will change to purple in those samples containing proteinaceous material. Record which samples give the purple color.

b. Xanthoproteic test

Two of the common amino acids found in proteins – tryptophan and tyrosine – contain substituted benzene rings in their side chains. When treated with moderately concentrated nitric acid, a nitro group ($-NO_2$) can also be added to the benzene rings, which results in the formation of a yellow color in the protein ("xanthoproteic" means "yellow protein"). For example, if nitric acid is spilled on the skin, the proteins in the skin will turn yellow as the skin is destroyed by the acid.

Set up a 250-mL beaker containing approximately 100 mL of water on a wire gauze/ring stand. Add a boiling chip and heat the water until it is gently boiling.

Place about 10 drops of 1% albumin, nonfat milk, and 1% alanine solutions, and a small cube of gelatin in separate clean small test tubes.

Add 8 to 10 drops of 6 *M* nitric acid *(Caution!)* to each test tube.

Carefully mix the contents of each test tube with a clean stirring rod, being sure to clean the stirring rod when switching between solutions. Make sure to break up the gelatin sample as much as possible when mixing.

Transfer the test tubes to the boiling water bath and heat the samples for three to four minutes.

Record which samples produce the yellow color characteristic of a positive xanthoproteic test. Which samples do not give a positive test?

c. Lead acetate test for sulfur

In strongly basic solutions, the amino acid cysteine from proteins will react with lead acetate solution to produce a black precipitate of lead sulfide, PbS. Cysteine is found in many (but not all) proteins. The strongly basic conditions are necessary to break up polypeptide chains into individual amino acids (hydrolysis).

Set up a 250-mL beaker containing approximately 100 mL of water on a wire gauze/ring stand and heat to boiling.

Place about 10 drops of 1% albumin, nonfat milk, and 1% alanine solutions, along with a small cube of gelatin, to separate clean, small, test tubes.

If you are willing to make the sacrifice, you may also test small samples of your hair (roll up 2 or 3 strands into a ball that can be fit in a test tube) or some fingernail clippings.

Add 15 to 20 drops of 6 *M* sodium hydroxide *(Caution!)* to each sample.

Carefully mix the contents of each test tube with a clean stirring rod, being sure to clean the stirring rod when switching between solutions. Make sure to break up the gelatin sample as much as possible when mixing.

Transfer the test tubes to the boiling water bath and heat the samples for 10 to 15 minutes. If the volume of liquid inside the test tubes begins to decrease during the heating period, add five to 10 drops of water to the test tubes to replace the volume.

After the heating period, allow the samples to cool. Then add five drops of 5% lead acetate solution to each test tube.

Record which samples produce a black precipitate of lead sulfide.

2. Denaturation of Proteins

Place 10 drops of 1% albumin solution in each of seven clean small test tubes. Set aside one sample as a control for comparisons.

Heat one albumin sample in a boiling water bath for five minutes and describe any change in the appearance of the solution after heating.

To the remaining albumin samples, add three or four drops of the following reagents (each to a separate albumin sample): 3 *M* hydrochloric acid; 6 *M* sodium hydroxide; saturated (5.4 *M*) sodium chloride solution; ethyl alcohol; 5 percent lead acetate solution.

Record any changes in the appearance of each albumin sample after adding the other reagent.

Repeat the tests above, using 10-drop samples of nonfat milk in place of the albumin solution.

EXPERIMENT 33

Proteins

Pre-Laboratory Questions

1. Draw the structures of any two of the common amino acids.

2. Using your two amino acids drawn above, draw structures for the two non-equivalent dipeptides that these amino acids could form.

3.   Explain what is meant by each of the different levels (primary, secondary, tertiary) of protein structure.

   a. primary

   _____

   _____

   _____

   _____

   _____

   b. secondary

   _____

   _____

   _____

   _____

   _____

   c. tertiary

   _____

   _____

   _____

   _____

   _____

4.   What are some factors that might cause the denaturing of a protein?

   _____

   _____

   _____

   _____

   _____

   _____

   _____

   _____

   _____

Name: _____ Section: _____

Lab Instructor: _____ Date: _____

EXPERIMENT 33

Proteins

Results/Observations

1. Tests for Protein

   Write below your observations of each of the tests on each of the protein samples.

   a. Biuret test

   albumin _____

   milk _____

   alanine _____

   gelatin _____

   b. Xanthoproteic test

   albumin _____

   milk _____

   alanine _____

   gelatin _____

   c. Lead Acetate test for sulfur

   albumin _____

   milk _____

   alanine _____

   gelatin _____

   hair _____

2. Denaturing of Protein

   albumin: Write your observations for each change made in the protein's environment.

   boiling _____

   HCl _____

   NaOH _____

   NaCl _____

   alcohol _____

   lead acetate _____

nonfat milk: Write your observations for each change made in the protein's environment.

boiling _____

HCl _____

NaOH _____

NaCl _____

alcohol _____

lead acetate _____

Questions

1.   What consequences might result if the proteins within a living cell were "denatured"? Can you think of any situations in which such denaturing might be advantageous?

_____

_____

_____

_____

_____

2.   You tested for the presence of sulfur in some proteins (the lead acetate test) and were invited to try testing a sample of your hair for the presence of sulfur. Use your textbook to describe how the presence of sulfur in hair protein makes "permanent waves" (perms) possible.

_____

_____

_____

_____

_____

_____

_____

_____

3.   The biuret test (Part 1a) and the xanthoproteic text (Part 1b) should have been negative for the alanine solution. Explain why.

_____

_____

_____

_____

_____

EXPERIMENT 34

Enzymes

Objective

Enzymes are the essential catalysts in biological systems that make life possible. In this experiment, you will examine the catalytic properties of several common enzymes.

Introduction

Enzymes are proteinaceous materials that function in the body as catalysts to control the speed of biochemical processes. Enzymes are said to be *specific* since a given enzyme typically acts only on a single type of molecule (or functional group): the molecule upon which an enzyme acts is referred to, in general, as the enzyme's *substrate*. Like other proteins, enzymes can be denatured by changes in their environment that cause the tertiary or secondary structures of the protein to deform.

The digestion of carbohydrates begins in the mouth, where the enzyme ptyalin is present in saliva. Ptyalin is a type of enzyme called an *amylase*, because the substrate it acts upon is the starch *amylose* (many enzymes are named after the substrate they act upon, with the ending of the substrate's name changed to *-ase* to indicate the enzyme). Ptyalin cleaves the long polysaccharide chain of amylose into smaller units called *oligosaccharides*. These oligosaccharides are then further digested into individual monosaccharide units in the small intestine.

Enzymes called *proteases* break down proteins into oligopeptides or free amino acids. For example, the protease called *bromelain* is present in fresh pineapple: if fresh pineapple is added to gelatin (a protein), the gelatin will not "set" because the protein is broken down. Meat tenderizers and several contact lens cleaners contain the enzyme *papain*, which is extracted from the papaya plant. Papain breaks down the proteins contained in muscle fibers of meat, making it more "tender," and it dissolves the protein deposits that may cloud contact lenses.

Hydrogen peroxide is often used to clean wounds because oxygen is released when H_2O_2 breaks down (this is observed as bubbling):

$$2H_2O_2 \rightarrow 2H_2O + O_2$$

Having an ample supply of oxygen available to a wound is thought to destroy or slow the growth of several dangerous anaerobic microorganisms (e.g., *Clostridium*), and the bubbling of the oxygen helps to cleanse the wound. Hydrogen peroxide used to clean a wound decomposes due to the action of enzymes called catalases (and similar enzymes called peroxidases). Such enzymes are also found in abundance in potatoes and yeasts.

Safety Precautions	<ul><li>Safety eyewear approved by your institution must be worn at all times while you are in the laboratory, whether or not you are working on an experiment.</li><li>Lead acetate solution is toxic if ingested. Wash hands after using. Dispose of as directed by the instructor.</li><li>Hydrogen peroxide solution may be corrosive to the eyes and skin. Wash immediately if spilled. Clean up all spills on the bench top.</li><li>The I_2/KI solution will stain skin and clothing if spilled.</li></ul>

Apparatus/Reagents Required

- 5% lead(II) acetate solution
- 0.1 M iodine/potassium iodide
- 6% hydrogen peroxide
- meat tenderizer
- contact lens cleaner
- fresh and canned pineapple samples
- potato
- yeast suspension
- oyster or soda crackers

Procedure

1. Action of Ptyalin (salivary amylase)

Heat approximately 100 mL of water in a 250-mL beaker to a gentle boil.

Using a small beaker, collect approximately 3 mL of saliva by letting the saliva flow freely from your mouth (do not spit).

Obtain a small soda or "oyster" cracker and crush it finely on a sheet of paper using the bottom of a small beaker.

Label four clean test tubes as 1, 2, 3 and 4. Transfer a small amount (about the size of a match head) of the crushed cracker into each test tube.

To the first test tube containing cracker add 2 mL of distilled water as a control.

To the second test tube containing cracker add 1 mL of your saliva and 1 mL of distilled water.

To the third test tube containing cracker add 1 mL of your saliva, followed by 1 mL of 5% lead acetate solution. As a heavy metal, the lead(II) ion should denature the protein of the salivary enzyme.

To the final test tube containing cracker, add 1 mL of your saliva and 1 mL of distilled water, and then place the test tube in the boiling water bath for 5-10 minutes. Heating the saliva sample should denature the protein of the salivary enzyme. Carefully remove the test tube from the boiling water bath and allow it to cool to room temperature.

Set the four test tubes aside on the bench top for at least one hour to allow the salivary enzyme to digest the starches in the cracker. Go on to the other portions of the experiment during the waiting period.

After the test tubes have stood for at least one hour, add one drop of 0.1 M I_2/KI solution to each of the test tubes. Iodine is a standard laboratory test for the presence of starch: if starch is present even in trace quantities in a sample, addition of I_2/KI solution will cause a deep blue/black color to appear. Record and explain your observations.

2. Action of Proteases

Your instructor will provide you with samples of prepared gelatin (a protein), either cast in the wells of a spot plate, or as small chunks that you can transfer to small test tubes. Although the gelatin is similar to the type used as a dessert, this gelatin has been made up double-strength to make it drier and easier to handle.

To one gelatin sample, add a few crystals of meat tenderizer.

To a second gelatin sample, add a few drops of the available commercial contact lens cleaner (record the brand and the name of the enzyme it contains, if available).

To a third gelatin sample, add a small piece of fresh pineapple (or juice extracted from fresh pineapple).

To a fourth gelatin sample, add a piece of cooked or canned pineapple.

Allow the gelatin samples to stand undisturbed for 30 to 60 minutes. Go on to the last part of the experiment while waiting.

After this time period, examine the gelatin samples for any evidence of breakdown of the gelatin by the proteases. Determine in particular if the gelatin has seemed to melt or partially melt. Record your observations.

3. Action of Catalase

Obtain two small slivers of freshly cut potato. Place one of the pieces of potato in a small test tube and transfer to a boiling water bath for 5 to 10 minutes.

Place 10 drops of 6% hydrogen peroxide (*Caution!*) into each of three small test tubes.

To one test tube, add eight to 10 drops of yeast suspension. Record your observations.

To the second test tube, add the small piece of uncooked, freshly-cut potato. Record your observations.

To the third test tube, add the small piece of cooked potato. Why does the cooked potato not cause the hydrogen peroxide to decompose.

Name: _____ Section: _____

Lab Instructor: _____ Date: _____

EXPERIMENT 34

Enzymes

Pre-Laboratory Questions

1.   Use your textbook or an encyclopedia to define or explain each of the following:

     a.   *enzyme*

     _____

     _____

     _____

     _____

     _____

     _____

     b.   *substrate*

     _____

     _____

     _____

     _____

     _____

     c.   *the "lock and key" model for enzymes*

     _____

     _____

     _____

     _____

     _____

     _____

d.   *what it means to say an enzyme is specific*

_____

_____

_____

_____

_____

_____

_____

e.   *what it means to say an enzyme has been denatured*

_____

_____

_____

_____

_____

_____

_____

EXPERIMENT 34

Enzymes

Results/Observations

1. Action of Ptyalin

    Results/explanation of Iodine test on cracker samples:

    Tube 1 (distilled water)

    _____

    _____

    Tube 2 (saliva/water)

    _____

    _____

    Tube 3 (saliva/water/lead)

    _____

    _____

    Tube 4 (saliva/water/heat)

    _____

    _____

2. Action of Proteases

    Observations of gelatin samples when treated with the proteolytic enzymes:

    meat tenderizer

    _____

    _____

    contact lens cleaner

    _____

    _____

    fresh pineapple

    _____

    _____

cooked/canned pineapple

_____

_____

3. Action of Catalase

    Observations on hydrogen peroxide solution when treated with catalase sources:

    yeast suspension

    _____

    _____

    uncooked potato

    _____

    _____

    cooked potato

    _____

    _____

Question

Several of the processes performed in this experiment were intended to show how a change in an enzyme's environment can denature the enzyme's protein, making the enzyme inactive towards the process it would ordinarily catalyze. Describe each of the denaturing processes performed, and tell how you realized that a given process had destroyed the enzyme's catalytic properties.

_____

_____

_____

_____

_____

_____

_____

_____

_____

_____

_____

_____

_____

Comparison of Antacid Tablets

Objective

Antacid tablets consist of weakly basic substances that are capable of reacting with the hydrochloric acid found in the stomach, converting the stomach acid into neutral or nearly neutral salts. In this experiment, you will determine the amount of stomach acid that two commonly used antacids are capable of neutralizing.

Introduction

In this experiment, you will attempt to evaluate the effectiveness in neutralizing hydrochloric acid of various commercial antacids. You will compare these products to the effectiveness of simple baking soda (sodium bicarbonate). The antacids will be dissolved in an excess of 0.5 M hydrochloric acid, and then the remaining acid (i.e., the portion of the acid that did not react with the antacid) will be titrated with standard sodium hydroxide solution.

It is not possible to titrate antacid tablets directly for several reasons. First, commercial antacid tablets frequently contain binders, fillers, flavorings and coloring agents that may interfere with the titration. Second, the bases found in most antacids are weak and become buffered as they are titrated, often leading to an indistinct indicator endpoint. Many of the weak bases used in antacid tables are also insoluble in water.

Your instructor may ask you to make a cost-effectiveness comparison of the various antacid brands.

Safety Precautions	• **Safety eyewear approved by your institution must be worn at all times while you are in the laboratory, whether or not you are working on an experiment.**
	• **The hydrochloric acid and sodium hydroxide solutions may be irritating to skin. Wash if these are spilled.**

Apparatus/Reagents Required

- buret and clamp
- 100-mL graduated cylinder
- plastic wrap
- two brands of commercial antacid tablets
- sodium bicarbonate
- phenolphthalein indicator solution
- standard 0.5 M hydrochloric acid and 0.5 M sodium hydroxide solutions

Procedure

Record all data and observations directly on the report page in ink.

Clean and rinse a buret with water. Then rinse and fill the buret with the available standard 0.5 M sodium hydroxide solution (record the exact concentration).

Obtain an antacid tablet: record the brand, the active ingredient, the amount of active ingredient, the price and the number of tablets per bottle (if available). Wrap the tablet in a piece of plastic wrap or notebook paper. Crush the tablet (use the bottom of a beaker or a heavy object). Transfer the crushed tablet to a clean 250-mL Erlenmeyer flask.

Obtain about 300 mL of the standard 0.5 M hydrochloric acid (record the exact concentration). With a graduated cylinder, measure exactly 100. mL of the hydrochloric acid and add it to the Erlenmeyer flask containing the crushed antacid tablet. Swirl the flask to dissolve the tablet as much as possible (the tablet is still dissolving if it is fizzing). *Note*: As mentioned earlier, commercial antacid tablets contain various other ingredients, which may not completely dissolve.

Add two to five drops of phenolphthalein indicator solution to the sample. The solution should not change color when the indicator is added, indicating that the solution is acidic (excess HCl).

If the sample turns pink/red when the phenolphthalein is added, this means that not enough hydrochloric acid was added to react completely with the antacid tablet and the solution is consequently basic. If the solution is pink/red, add 0.5 M HCl in 10-mL increments (record the amount used) until the sample is colorless. Record the total volume of HCl added.

Record the initial reading of the buret. Titrate the antacid sample with standard NaOH solution until the solution just barely turns pink. Record the final reading of the buret at the color change.

Repeat the procedure two more times using the same brand of antacid tablet.

Repeat the procedure either with another brand of commercial antacid tablet (three samples) or with samples of baking soda (sodium bicarbonate) on the order of 0.7 g (record the exact weight used). If sodium bicarbonate is used, add the standard 0.5 M HCl to it very slowly to prevent excessive frothing as carbon dioxide is liberated.

Calculations

Example: Suppose an antacid tablet was dissolved as described above in 100. mL of 0.501 M HCl, and then was titrated with standard 0.498 M NaOH to a phenolphthalein endpoint, requiring 26.29 mL of the NaOH solution.

$$\text{millimol HCl taken} = (100.\text{ mL}) \times \left(\frac{0.501 \text{ mmol}}{1 \text{ mL}}\right) = 50.1 \text{ mmol HCl}$$

$$\text{millimol NaOH (HCl not consumed)} = (26.29 \text{ mL})\left(\frac{0.498 \text{ mmol}}{1 \text{ mL}}\right) = 13.1 \text{ mmol}$$

$$\text{millimol HCl neutralized} = 50.1 \text{ mmol} - 13.1 \text{ mmol} = 37.0 \text{ mmol HCl consumed}$$

Given that the molar mass of HCl is 36.46 g, calculate the mass of HCl consumed by the tablet

$$(37.0 \text{ mmol HCl}) \times \left(\frac{36.46 \text{ mg HCl}}{1 \text{ mmol HCl}}\right) = 1349 \text{ mg} = 1.35 \text{ g HCl consumed by the tablet}$$

Cost Effectiveness

There often is a large difference in price among the various brand name and generic or store brand ant-acids. Your instructor may ask you to make a comparison among the different brands of antacid used for this experiment if the pricing information is available. To compare the cost effectiveness of one antacid versus another, you need to determine the cost of the antacid per gram of HCl consumed. Consider these two antacids:

Brand	No. of tablets in bottle	Price	g HCl consumed/tablet
Brand A	100	$3.99	0.789 g
Brand B	30	$4.99	1.523 g

$$\text{For Brand A: Cost} = \frac{\$3.99}{(100\,\text{tablets})(0.789\,\text{g/tablet})} = \$0.0506 \text{ per gram of HCl consumed}$$

$$\text{For Brand B: Cost} = \frac{\$4.99}{(30\,\text{tablets})(1.523\,\text{g/tablet})} = \$0.109 \text{ per gram of HCl consumed}$$

So although Brand B consumes much more HCl per tablet, it costs nearly twice as much per gram of HCl consumed because there are fewer tablets in the bottle. If Brand A and Brand B were found to contain the same active ingredient, then Brand A is a much better buy.

EXPERIMENT 35

Comparison of Antacid Tablets

Pre-Laboratory Questions

1. What is a standard solution of acid or base?

   _____

   _____

   _____

2. Some of the common bases used as the active ingredients in commercial antacid tablets are listed below. Calculate the number of milliliters of 0.500 M HCl solution that could be neutralized by 1.00 g of each of these substances. Show your calculations.

   $CaCO_3$

   $NaHCO_3$

   $Mg(OH)_2$

   $Al(OH)_3$

3. In this experiment, rather than directly titrating the bases in the antacid tablet with a standard hydrochloric acid solution, you first dissolve the tablet in an excess of standard hydrochloric acid solution, and then titrate with standard sodium hydroxide the portion of hydrochloric acid that is not consumed by the tablet. Why is this indirect approach necessary?

_____

_____

_____

_____

_____

_____

_____

_____

Name: _____ Section: _____

Lab Instructor: _____ Date: _____

EXPERIMENT 35

Comparison of Antacid Tablets

Results/Observations

Concentrations of standard solutions:

HCl _____ NaOH _____

First antacid: Brand _____

Label information: _____

	Sample 1	Sample 2	Sample 3
Mass of tablet used	_____	_____	_____
Volume of 0.5 M HCl used	_____	_____	_____
Initial reading of buret	_____	_____	_____
Final reading of buret	_____	_____	_____
mL of 0.5 M NaOH used	_____	_____	_____
Mass of HCl consumed by tablet	_____	_____	_____
Mean mass of HCl consumed by tablet	_____		

Second antacid: Brand _____

Label information _____

	Sample 1	Sample 2	Sample 3
Mass of tablet used	_____	_____	_____
Volume of 0.5 M HCl used	_____	_____	_____
Initial reading of buret	_____	_____	_____
Final reading of buret	_____	_____	_____
mL of 0.5 M NaOH used	_____	_____	_____
Mass of HCl consumed by tablet	_____	_____	_____
Mean mass of HCl consumed by tablet	_____		

Questions

1. Which of the two antacids tested consumed more HCl per gram? Which consumed more HCl per tablet? If the prices of the antacids are available, which antacid is a better buy?

_____

_____

_____

_____

_____

_____

2. Generally, the substances used as antacids are either weak bases or very insoluble bases. Why is a strong soluble base such as NaOH not used in antacid tablets?

_____

_____

_____

_____

3. Read the label(s) on the commercial antacid tablets you used. List the brand name(s) and the active antacid ingredients.

_____

_____

_____

_____

_____

_____

4. Some people overuse antacids. Consult a chemical encyclopedia to find out what side effects may occur if antacids are used too frequently.

_____

_____

_____

_____

_____

_____

EXPERIMENT 36

Determination of Vitamin C in Fruit Juices

Objective

The concentration of Vitamin C in commercially available fruit juices will be determined.

Introduction

Vitamin C is an important essential anti-oxidant found in fresh fruits and vegetables.

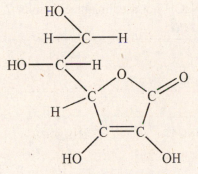

Vitamin C (L-ascorbic acid)

Vitamin C is necessary in humans for the production of the protein collagen and as co-factor in several enzyme processes. Although many animals can synthesize Vitamin C, humans lack a necessary enzyme for its synthesis and have to consume Vitamin C in their diet. Lack of Vitamin C results in the disease condition called scurvy, which causes spots on the skin, gum disease, loss of teeth and bleeding. Before the advent of steamships, sailors on long ocean voyages in sailing ships frequently developed scurvy because of a lack of fresh fruit and vegetables in their diet. Vitamin C is destroyed by oxygen, light and heat and so only fresh fruits and vegetables contain significant amounts.

In this experiment, you will determine the amount of Vitamin C present in a fruit juice by titration with an iodine/potassium iodide solution. Iodine in the presence of potassium iodide forms the triiodide ion, I_3^-. The triiodide ion is able to oxidize the Vitamin C molecule quantitatively, forming dehydroascorbic acid.

$$C_6H_8O_6 + I_3^- + H_2O \rightarrow C_6H_6O_6 + 3I^- + 2H^+$$
$$\text{ascorbic acid} \qquad\qquad\qquad \text{dehydroascorbic acid}$$

You will determine the concentration of the iodine solution by reacting it first with a standard solution containing a known amount of Vitamin C per milliliter. The concentration of the iodine solution will be determined in terms of what we will call the "Vitamin C equivalency factor." This factor will represent how many mg of Vitamin C each milliliter of the iodine solution is able to oxidize. Use of this factor means that we can titrate multiple samples of fruit juices with the iodine, and then just multiply the number of milliliters of iodine that are required in the titration by the equivalency factor to get the Vitamin C content of the fruit juice.

Safety Precautions	• Safety eyewear approved by your institution must be worn at all times while you are in the laboratory, whether or not you are working on an experiment. • The iodine solution will stain skin and clothing if spilled. Transfer the buret and stand to the floor before filling the buret with iodine solution. Use a funnel. • The fruit juices used in this experiment are to be treated as "chemicals" and may not under any circumstances be consumed or removed from the laboratory.

Apparatus/Reagents Required

- buret and clamp
- 25-mL pipet and safety bulb
- 250-mL volumetric flask
- 250-mL Erlenmeyer flasks
- assorted beakers
- funnel, watch glass
- Vitamin C tablets (250. mg Vitamin C per tablet)
- 1% starch solution
- iodine solution
- mortar and pestle
- fruit juice samples

Procedure

1. Preparation of the Vitamin C Standard Solution

Obtain a 250 mL volumetric flask and clean it out with soap and water. Rinse with several portions of tap water. Rinse with several portions of deionized or distilled water. Examine the flask carefully and note the location of the calibration mark on the neck of the flask. When the flask is filled exactly to the calibration mark, the flask will contain 250.0 mL.

The neck of the flask above the calibration mark must be especially clean so that water does not bead up above the calibration mark during the preparation of the solution. If water beads up in this region, scrub the neck of the flask with a brush, rinse, and check again to see if the flask is clean. Any water retained above the calibration mark will run into the solution being prepared and change the volume of the solution. Make sure the volumetric flask is rinsed several times with distilled or deionized water prior to use.

Obtain a 250-mg Vitamin C tablet. In this experiment, we will assume that manufacturing specifications at the company that manufactures the Vitamin C are very rigid, and that the tablet contains exactly 250.

mg of Vitamin C. Vitamin C tablets are typically fairly hard because of binding agents that have been added, and they dissolve slowly unless they are first crushed. Use a mortar and pestle to grind the tablet into a uniform powder.

Place a funnel into the neck of the volumetric flask, and quantitatively transfer the crushed Vitamin C tablet from the mortar, through the funnel, into the volumetric flask. If any of the tablet adheres to the mortar, use a stream of water from your plastic wash bottle to try to wash the tablet particles into the funnel. Once all the tablet has been transferred to the volumetric flask, add water to the flask about half-way up the long neck of the flask: do not add water all the way to the calibration mark at this point.

Stopper the volumetric flask and invert and shake it to dissolve the Vitamin C tablet. Continue shaking until all pieces of the tablet have dissolved. Then use a plastic dropper to add water to exactly the calibration mark of the flask.

Given that you dissolved a 250. mg Vitamin C tablet to a final volume of 250.0 mL, calculate the concentration of the standard Vitamin C solution in terms of how many milligrams of Vitamin C are contained per milliliter of the solution.

2. Preparation of the Buret

Obtain about 150 mL of the available iodine solution in a beaker. Keep the beaker covered with a watch glass to prevent evaporation.

The fruit juice analysis will be performed as a titration. In a titration, we prepare a sample in an Erlenmeyer flask of the material to be analyzed, and then determine with a buret the exact volume of a reagent required to react completely with the sample.

Obtain a 50-mL buret and clean it with soap and water until water runs in sheets down the inside of the buret without beading up anywhere on the walls of the buret. Set up the buret in a clamp on a ring stand. See Figure 36-1.

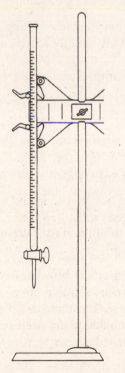

Figure 36-1. Set up for titration

Move the buret/stand to the floor and place a clean, dry funnel in the stem of the buret. The buret will be filled on the floor because the iodine solution to be placed into the buret can stain skin and clothing if spilled. Add iodine solution to the buret until the liquid level is at least an inch above the top of the scale of the buret.

Remove the funnel and transfer the buret and stand back to the lab bench. Place a beaker under the tip of the buret, open the stopcock of the buret, and run iodine solution through the tip of the buret until there are no air bubbles in the tip. If the liquid level in the buret at this point is not below the top of the scale of the buret, run out a little more iodine solution until the liquid level is just below the top of the scale.

3. Preparation of the Vitamin C Standard Samples

Label three 250-mL Erlenmeyer flasks as samples 1, 2 and 3, then clean out the flasks with soap solution, and finally rinse them with several portions of water. Perform a final rinse with several portions of deionized or distilled water.

Obtain a 25-mL volumetric pipet and rubber safety bulb. Using the bulb, draw a small amount of soap solution into the pipet (five to 10 mL). Then put your fingers over both ends of the pipet, hold the pipet horizontally, and rotate the barrel of the pipet so as to wash all inside surfaces of the pipet with the soap solution. Allow the soap solution to drain from the pipet.

Rinse the pipet at least three times to remove all traces of soap solution by drawing small portions of water into the pipet and rinsing the inside surfaces of the pipet as described above with deionized or distilled water. Allow the water to drain from the pipet before starting a new rinsing each time.

Transfer about 100 mL of your standard Vitamin C solution from the volumetric flask to a 250-mL beaker. Using the bulb, draw five to 10 mL of the Vitamin C solution into the pipet, then rinse the inside surfaces of the pipet with the Vitamin C solution as described above. This step is to make sure all rinse water has been removed from the pipet, so that we can pipet an exact amount of Vitamin C without diluting it. Repeat the rinsing with 5–10 mL portions of standard Vitamin C solution twice more.

Finally, pipet exactly 25.00 mL of the Vitamin C tablet solution into each of the three sample flasks. How many mg of Vitamin C does the 25.00 mL sample represent?

4. Titration of the Vitamin C Standard Samples

Add approximately 75 mL of deionized or distilled water to Sample 1, and then add approximately 1 mL of 1% starch solution. Swirl the flask to mix.

As the titration takes place, iodine from the solution in the buret will react with Vitamin C in the solution in the Erlenmeyer flask. When all the Vitamin C in the sample has reacted, the next drop of iodine solution added will be in excess. Iodine and starch react with each other to form an intensely colored blue/black complex. The appearance of a permanent blue-black color in the flask (one that does not fade on swirling) is taken as the point where the Vitamin C in the sample has completely reacted.

Take a reading of the initial level of iodine solution in the buret and record.

Place Sample 1 under the tip of the buret and add a few drops of iodine solution from the buret. As the iodine solution enters the Vitamin C solution, a dark blue (nearly black) cloud will form in the solution: a momentary excess of iodine occurs as the iodine solution enters the solution in the flask. Swirl the flask and note that the dark blue cloud disappears as the iodine mixes with more of the Vitamin C sample. The endpoint of the titration is the point at which adding one additional drop of iodine causes the dark blue color to remain even after the solution is swirled, signaling that all the Vitamin C has reacted.

Continue adding iodine slowly, with swirling of the flask. As you get closer and closer to the endpoint of the titration, it will take longer and longer for the dark blue color to dissipate. Begin adding iodine one drop at a time, with swirling, until one drop of iodine causes the blue color to remain even after the solution is swirled. If you are very careful, you can find the exact drop of iodine that causes the blue color to remain (the solution will only be pale blue, because there will only be a very slight excess of iodine).

Record the final liquid level in the buret.

Repeat the procedure for Samples 2 and 3.

Calculate the "Vitamin C equivalency factor" for your three samples and the average value of this factor.

> **Example**:
>
> A 20.00 mL sample of 2.00 mg/mL vitamin C solution required 31.75 mL of iodine solution to reach the blue starch endpoint. Calculate the Vitamin C equivalency factor for the iodine solution in terms of how many mg of Vitamin C each mL of the iodine solution is able to react with.
>
> $$\text{mg Vit. C} = 20.00 \text{ mL Vitamin C solution} \times \frac{2.00 \text{ mg Vitamin C}}{1 \text{ mL Vitamin C solution}} = 40.0 \text{ mg Vitamin C}$$
>
> $$\frac{40.0 \text{ mg Vitamin C}}{31.75 \text{ mL iodine solution}} = 1.26 \text{ mg Vitamin C/mL iodine solution}$$

> **Example:**
>
> Suppose a fruit juice sample was titrated with the iodine solution described above. If 25.00 mL of the fruit juice required 19.25 mL of iodine solution to reach the blue starch endpoint, how many milligrams of Vitamin C are contained in each milliliter of the fruit juice?
>
> $$19.25 \text{ mL iodine solution} \times \frac{1.26 \text{ mg Vitamin C}}{1 \text{ mL iodine solution}} = 24.3 \text{ mg Vitamin C.}$$
>
> This mass of Vitamin C was contained in 25.00 mL of the fruit juice. Therefore the number of milligrams of Vitamin C per milliliter of the fruit juice is
>
> $$\frac{24.3 \text{ mg Vitamin C}}{25.00 \text{ mL juice}} = 0.972 \text{ mg Vitamin C/mL juice}$$

5. Titration of Fruit Juice Samples

Your instructor will have available for you one or more freshly-opened samples of whole fruit juice. Take about 100 mL of one of the juices in a small beaker and record the name of the juice. If your juice sample has pulp in it, filter the juice either through filter paper or through a double-thickness of cheesecloth held in your gravity funnel.

Clean out the three Erlenmeyer flasks and pipet 25.00 mL of juice into each.

Refill your buret with iodine solution if needed and take a reading of the initial liquid level.

Add approximately 75 mL of deionized or distilled water to fruit juice Sample 1, and then add approximately 1 mL of 1% starch solution also. Swirl the flask to mix.

Titrate Sample 1 using the same technique as used for titrating the standard Vitamin C samples. Because some fruit juices are cloudy, the appearance of the endpoint may not be as sharp as it was for the standard Vitamin C samples. Try to find the point where one additional drop of iodine from the buret will cause a permanent blue color that does not fade on swirling.

Record the final liquid level in the buret.

Repeat the process for fruit juice Samples 2 and 3.

Using the average "Vitamin C equivalency factor" you determined for your standard Vitamin C samples, calculate the number of milligrams of Vitamin C present per milliliter of fruit juice.

Compare your results with those obtained by other students. Of the fruit juices available in the lab, which contains the most Vitamin C per milliliter?

6. Additional/Optional Titrations

Your instructor may ask you to titrate additional fruit juice samples so that you can make your own comparison of Vitamin C content.

You may also be asked to investigate how heating or exposure to air affects the Vitamin C level in fruit juices. For example, you might titrate a fruit juice that was left in an open container overnight, or that had been heated strongly (as during the canning process).

Regardless of the type of juice, or how the juice has been treated, the titration process is the same as you have performed for the fresh juice.

Name: _____ Section:_____

Lab Instructor: _____ Date:_____

EXPERIMENT 36

Determination of Vitamin C in Fruit Juices

Pre-Laboratory Questions

1. Given the equation for the oxidation of ascorbic acid (Vitamin C) to dehydroascorbic acid given in the Introduction, write the half-reactions for the oxidation and the reduction processes. How many electrons does each ascorbic acid molecule lose when it is oxidized?

    _____

    _____

    _____

    _____

    _____

    _____

    _____

    _____

    _____

    _____

2. Why is it important that the interior surface above the calibration mark of a volumetric flask be absolutely clean? What error would be introduced in the concentration of a solution if water beaded up above the calibration mark? Would the solution be more concentrated or less concentrated than expected?

    _____

    _____

    _____

    _____

    _____

    _____

    _____

    _____

3. Suppose a 25.00 mL pipet sample of 1.50 mg/mL ascorbic acid solution was titrated with an iodine solution, requiring 31.75 mL of the iodine solution to reach the blue-black starch endpoint. Suppose this same iodine solution was then used to titrate 25.00 mL of fruit juice, requiring 29.34 mL to reach the endpoint. What is the concentration of ascorbic acid in the fruit juice sample in mg/mL?

    _____

    _____

    _____

    _____

    _____

    _____

    _____

4. Use the Internet or a chemical encyclopedia to list five foods that are especially good sources of Vitamin C.

    _____

    _____

    _____

    _____

    _____

5. Vitamin C is a very popular vitamin. Many claims are made about it in the popular press. Use the Internet or a chemical encyclopedia to list five conditions or processes in which Vitamin C is purported to be beneficial.

    _____

    _____

    _____

    _____

    _____

    _____

    _____

    _____

    _____

    _____

    _____

    _____

EXPERIMENT 36

Determination of Vitamin C in Fruit Juices

Results/Observations

Concentration of standard Vitamin C solutions _____

Titration of the Vitamin C Standard Samples (Part 4)

	Sample 1	Sample 2	Sample 3
Initial reading of buret	_____	_____	_____
Final reading of buret	_____	_____	_____
mL of iodine solution used	_____	_____	_____
Vitamin C equivalency factor	_____	_____	_____
Mean Vitamin C equivalency factor	_____		

Titration of Fruit Juice Samples (Part 5)

Which fruit juice did you use? _____

	Sample 1	Sample 2	Sample 3
Initial reading of buret	_____	_____	_____
Final reading of buret	_____	_____	_____
mL of iodine solution used	_____	_____	_____
mg Vitamin C/mL juice	_____	_____	_____
Mean mg Vitamin C/mL juice	_____		

Titration of Additional Samples (Part 6)

Which additional juice sample did you use? _____

_____

	Sample 1	Sample 2	Sample 3
Initial reading of buret	_____	_____	_____
Final reading of buret	_____	_____	_____
mL of iodine solution used	_____	_____	_____
mg Vitamin C/mL juice	_____	_____	_____
Mean mg Vitamin C/mL juice	_____		

Questions

1. Show how you calculated the Vitamin C equivalency factor for your Sample 1 standard.

    _____

    _____

    _____

    _____

    _____

    _____

    _____

    _____

2. Of the fruit juices available in the lab, which had the highest Vitamin C content per milliliter? Which contained the least Vitamin C per milliliter?

    _____

    _____

    _____

    _____

    _____

3. If you performed an additional titration involving a juice that had been left open to the air or that had been heated, how did the Vitamin C content of that sample compare to the Vitamin C content in the fresh juices? Explain.

    _____

    _____

    _____

    _____

    _____

4. In addition to protecting against scurvy, Vitamin C is recommended in the diet because it is an anti-oxidant. Use the Internet or a chemical encyclopedia to define anti-oxidant.

    _____

    _____

    _____

    _____

APPENDIX A

Significant Figures

Indicating Precision in Measurements

When a measurement is made in the laboratory, it is generally made to the limit of precision permitted by the measuring device used. For example, when a very careful determination of mass is needed, an analytical balance is used, which permits the determination of mass to the nearest 0.0001 g. If such a balance were used in an experiment, it would be silly to record the mass determined for an object as "three grams" if the balance permits a mass determination such as 3.1123 g. It would be just as silly, however, to record your weight as determined on a bathroom scale to the nearest 0.0001 pound. Bathroom scales simply do not justify that much precision.

To indicate the actual precision with which a measurement has been made, scientists are very careful about how many *significant figures* they report in their results and data. The significant figures in a number consist of all the digits in the number whose values are known with complete certainty, plus a final digit that is a best estimate of the reading between the smallest scale divisions on the instrument used to measure the number.

For example, the thermometer in a student chemistry locker has scale divisions marked on the barrel of the thermometer for every whole degree from –20°C to 110°C. Suppose the mercury level in the thermometer were approximately halfway between the 35° and 36° marks. The temperature indicated by the thermometer would be known with complete certainty to be at least 35°. If the student using the thermometer made an estimate that the mercury level was 0.4 of the way from 35° to 36°, then the temperature recorded by the student should be 35.4°C, with the digit in the first decimal place a best estimate of the reading between the smallest scale divisions of the thermometer. The temperature would be reported correctly to three significant figures. Any scientist seeing the temperature 35.4° would realize that a thermometer with a scale divided to the nearest whole degree had been used for the measurement.

To help you, since you are just beginning your study of chemistry, this manual tries, in most instances, to point out the precision required for an experiment. For example, a procedure may say to "weigh 25 g of salt to the nearest 0.01 g." This statement indicates both how much of the substance to use (25 g) and the precision necessary for the experiment to have meaningful results. (Because the mass is to be measured to the nearest 0.01 g, the mass is needed to four-significant-figure precision – two figures before and two figures after the decimal point.)

A point that often confuses beginning chemistry students is the meaning of a zero in a number. Sometimes zeros are significant figures, and sometimes they are not. If a zero is *embedded* in a number, the zero is always significant and indicates that the reading of the scale of the instrument used for the measurement actually was zero.

Examples:

      1.203 g   10.01 mL

When zeros occur at the beginning of a number, they are never significant. Zeros at the beginning of a number are just indicating the position of the decimal point relative to the first significant digit.

Examples:

0.25 g   **0.00**1 L

One way to show that initial zeros are not significant is to write numbers in scientific notation. For example, the previous numbers could be written as

2.5×10^{-1} g 1×10^{-3} L

When numbers with initial zeros are written in this manner, the zeros are not even indicated.

Zeros at the end of numbers may or may not be significant, depending on the actual measurement indicated. For example, in the relationship

1 L = **1,000** mL

the three zeros in 1,000 mL are significant. They indicate that one liter is defined to be precisely 1,000 milliliters. However, if you reported the weight of a beaker as 20.0000 g, your instructor would probably question your result. It is very unlikely that the weight of a beaker would have this precise weight.

To avoid confusion about whether or not zeros at the end of a number are significant, the scientist again resorts to scientific notation. If the weight of 20.0000 g were exactly precise, this could be indicated by writing the mass as

2.00000×10^{1} g

When trailing zeros are written in scientific notation, the zeros are assumed to be significant and to represent actual scale readings on the measuring device.

Arithmetic and Significant Figures

Although hand-held calculators have taken much of the drudgery out of mathematical manipulations in science, the use of such calculators has led to a whole new sort of problem. Calculators generally are able to handle at least eight digits in their displays; calculators assume that all numbers punched in have this number of significant figures and give all answers with this number of digits. For example, if you enter the simple problem

2 ÷ 3 = ?

a calculator will respond with an answer of 0.6666667. Clearly, if two and three represent measurements in the laboratory that were made only to the nearest whole number, then a quotient like 0.6666667 implies too much precision. Not all the digits in the answer indicated by the calculator are significant.

A. Addition and Subtraction

When adding and subtracting numbers that are known to different levels of precision, we should record the answer only to the number of decimal places indicated by the number known to the least number of decimal places. Consider this addition:

20.2354 + 1.02 + 337.114

Although 20.2354 is significant to the fourth decimal place, and 337.114 to the third decimal place, the number 1.02 is known only to the second decimal place. The sum of these numbers then can be reported correctly only to the second decimal place. Perform the arithmetic, but then round the answer so that it extends only to the second decimal place:

$$20.2354 + 1.02 + 337.114 = 358.3694 \text{ which should be reported as } 358.37.$$

B. Multiplication and Division

When we multiply or divide numbers of different precision, the product or quotient should have only as many digits reported as the least precise number involved in the calculation. For example, consider the multiplication

$$(3.2795)(4.3302)(2.1)$$

Both 3.2795 and 4.3302 are known to five significant figures, whereas 2.1 is known with considerably less precision (to only two significant figures). The product of this multiplication may be reported only to two significant figures:

$$(3.2795)(4.3302)(2.1) = 29.821871 \text{ (calculator)} = 30.$$

Even though a calculator will report 29.821871 as the product, this answer must be rounded to two significant figures to indicate the fact that one of the numbers used in the calculation was known with a lower level of precision.

APPENDIX B

Exponential Notation

Writing Large and Small Numbers in Exponential Notation

Many of the numbers used in science are very large or very small and are not conveniently written in the sort of notation used for normal-sized numbers. For example, the number of atoms of carbon present in 12.0 g of carbon is

602,000,000,000,000,000,000,000

Clearly, in a number of this magnitude, most of the zeros indicated are merely place-holders used to locate the decimal point in the correct place.

With very large and very small numbers, it is usually most convenient to express the number in scientific or exponential notation. For example, the number of atoms in 12 g of carbon could be written in exponential notation as

6.02×10^{23}

This expression indicates that 6.02 is to be multiplied by ten, 23 times. To see how to convert a number to exponential notation, consider the following example.

Example:

Write 125,000 in exponential notation.

The standard format for exponential notation uses a factor between one and 10 that gives all the appropriate significant digits of the number, multiplied by a power of 10 that locates the decimal point and indicates the *order of magnitude* of the number.

The number 125,000 can be thought of as $1.25 \times 100,000$.

100,000 is equivalent to $10 \times 10 \times 10 \times 10 \times 10 = 10^5$.

125,000 in exponential notation is then 1.25×10^5.

A simple method for converting a large number to exponential notation is as follows. Move the decimal point of the given number to the position after the first significant digit of the number. This placement gives the multiplying factor for the number when it has been written in exponential notation. Count the number of places the decimal point has been moved from its original position (at the end of the number) to its new position (after the first significant digit). The number of places the decimal point has been moved represents the exponent of the power of 10 of the number when expressed in scientific notation.

Example:

Write the following in exponential notation: 1,723

Move decimal point to position after the first significant digit: 1.723

Decimal point has been moved three spaces to the left from the original 1,723 to give 1.723.

Exponent will be three.

$1,723 = 1.723 \times 10^3$

Example:

Write the following in exponential notation: 7,230,000

Move decimal point to position after the first significant digit: 7.230000

Decimal point has been moved six spaces to the left from its original position.

Exponent will be six.

$7,230,000 = 7.23 \times 10^6$

The same methods are used when attempting to express very small numbers in exponential notation. For example, consider this number:

0.0000359

This number is equivalent to 3.59×0.00001, and because 0.00001 is equivalent to 10^{-5},

$0.0000359 = 3.59 \times 10^{-5}$

A simple method for converting a small number to exponential notation is as follows. Move the decimal point of the given number to the position after the first significant digit of the number. Now you have the multiplying factor for the number when it has been written in exponential notation. Count the number of places the decimal point has been moved from its original position (at the beginning of the number) to its new position (after the first significant digit). The number of places the decimal point has been moved represents the exponent of the power of 10 of the number when expressed in scientific notation. Because the number is smaller than one, the exponent is negative.

Example:

Write the following in exponential notation: 0.0000000072

Move decimal point to position after the first significant digit: 7.2

Decimal point has been moved nine spaces from its original position at the beginning of the number to after the first significant digit. Exponent will thus be –9 (negative nine):

$0.0000000072 = 7.2 \times 10^{-9}$

Example:

Write the following in exponential notation: 0.00498

Move decimal point to position after first significant digit: 4.98

Decimal point has been moved three spaces from its original location. Exponent will be –3:

$0.00498 = 4.98 \times 10^{-3}$

Arithmetic with Exponential Numbers

In chemistry problems it often becomes necessary to perform arithmetical operations with numbers written in exponential notation. Such arithmetic is basically no different from arithmetic with normal numbers, but certain methods for handling the exponential portion of the numbers are necessary.

A. Addition and Subtraction of Exponential Numbers

Numbers written in exponential notation can be added together or subtracted only if they have the same power of 10. Consider this problem:

$$1.233 \times 10^3 + 2.67 \times 10^2 + 4.8 \times 10^1$$

It is not possible to just add up the coefficients of these numbers. This problem, if written in normal (non-exponential) notation would be:

$$1{,}233 + 267 + 48 = 1{,}548$$

Before numbers written in exponential notation can be added, they must all be converted to the same power of 10. For example, the numbers could all be written so as to have 10^3 as their power.

$$1{,}233 = 1.233 \times 10^3$$

$$2.67 \times 10^2 = 0.267 \times 10^3$$

$$4.8 \times 10^1 = 0.048 \times 10^3$$

and the sum as $(1.233 + 0.267 + 0.048) \times 10^3 = 1.548 \times 10^3 = 1{,}548$.

B. Multiplication and Division of Exponential Numbers

According to the commutative law of mathematics, the expression $(A \times B)(C \times D)$ could be written validly as $(A \times C)(B \times D)$. When this law is applied to numbers written in exponential notation, it becomes very straightforward to find the product of two such numbers.

Example:

Evaluate: $(4.2 \times 10^6)(1.5 \times 10^4)$

$(4.2 \times 10^6)(1.5 \times 10^4) = (4.2 \times 1.5)(10^6 \times 10^4)$

$(4.2 \times 1.5) = 6.3$

$(10^6 \times 10^4) = 10^{(6+4)} = 10^{10}$

$(4.2 \times 10^6)(1.5 \times 10^4) = 6.3 \times 10^{10}$

When numbers written in exponential notation are divided, a similar application of the commutative law of mathematics is made: $(E \times F)/(G \times H) = (E/G) \times (F/H)$

Example:

Evaluate: $(4.8 \times 10^9)/(1.5 \times 10^7)$

$(4.8 \times 10^9)/(1.5 \times 10^7) = (4.8/1.5) \times (10^9/10^7)$

$(4.8/1.5) = 3.2$

$(10^9/10^7) = 10^{(9-7)} = 10^2$

$(4.8 \times 10^9)/(1.5 \times 10^7) = 3.2 \times 10^2$

APPENDIX C

Plotting Graphs in the Introductory Chemistry Laboratory

Very often the data collected in the chemistry laboratory are best presented pictorially, by means of a simple graph. This section reviews how graphs are most commonly constructed, but you should consult with your instructor about any special considerations he or she may wish you to include in the graphs you will be plotting in the laboratory.

The use of a graphical presentation allows you to display clearly both experimental data and any relationships that may exist among the data. Graphs also permit interpolation and extrapolation of data, to cover those situations that were not directly investigated in an experiment and to allow predictions to be made for such situations. In order to be meaningful, your graphs must be well planned. Study the data before attempting to plot them. Recording all data in tabular form in your notebook will be a great help in organizing the data.

The particular sort of graph paper you use for your graphs will differ from experiment to experiment, depending on the precision possible in a particular situation. For example, it would be foolish to use graph paper with half-inch squares for plotting data that had been recorded to four-significant-figure precision. It is impossible to plot such precise data on paper with such a large grid system. Similarly, it would not be correct to plot very roughly determined data on very fine graph paper. The fine scale would imply too much precision in the determinations.

A major problem in plotting graphs is deciding what each scale division on the axes of the graph should represent. You must learn to scale your data, so that the graphs you construct will fill virtually the entire graph paper page. Scale divisions on graphs should always be spaced at very regular intervals, usually in units of 10, 100, or some other appropriate power of 10. Each grid line on the graph paper should represent some readily evident number (for example, not 1/3 or 1/4 unit). It is not necessary or desirable to have the intersection of the horizontal and vertical axes on your graph always represent the origin (0, 0). For example, if you were plotting temperatures from 100°C to 200°C, it would be silly to start the graph at zero degrees. The axes should be labeled in ink as to what they indicate, and the scale divisions should be clearly marked. By studying your data, you should be able to come up with realistic minimum and maximum limits for the scale of the graph and for what each grid line on the graph paper will represent. Consult the many graphs in your textbook for examples of properly constructed graphs.

By convention, the horizontal (x) axis of a graph is used for plotting the quantity that the experimenter has varied during an experiment (independent variable). The vertical (y) axis is used for plotting the quantity that is being measured in the experiment (as a function of the other variable). For example, if you were to perform an experiment in which you measured the pressure of a gas sample as its temperature was varied, you would plot temperature on the horizontal axis, since this is the variable that is being controlled by the experimenter. The pressure, which is measured and results from the various temperatures used, would be plotted on the vertical axis.

When plotting the actual data points on your graph, use a sharp, hard pencil. Place a single, small, round dot to represent each datum. If more than one set of data is being plotted on the same graph, small squares or triangles may be used for the additional sets of data. If an estimate can be made of the probable magnitude of error in the experimental measurements, indicate it with *error bars* above and below each data point (the size of the error bars on the vertical scale of the graph should indicate the

Appendix C

magnitude of the error on the scale). To show the relationship between the data points, draw the best possible straight line or smooth curve through the data points. Straight lines should obviously be inked in with a ruler, whereas curves should be sketched using a drafter's French curve.

If the data plot for an experiment gives rise to a straight line, you may be asked to calculate the *slope* and *intercept* of the line. Although the best way to determine the slope of a straight line is by the techniques of numerical regression, if the data seem reasonably linear, the slope may be approximated by:

$$\text{slope} = (y_2 - y_1)/(x_2 - x_1)$$

Here, (x_1, y_1) and (x_2, y_2) are two points on the line. Once the slope has been determined, either intercept of the line may be determined by setting y_1 or x_1 equal to zero (as appropriate) in the equation of a straight line:

$$(y_2 - y_1) = (\text{slope})(x_2 - x_1)$$

If your university or college has microcomputers available for your use, simple computer programs for determining the slope and intercept of linear data are commonly available. Spreadsheet programs also have the ability to create "x-y" graphs. Use of computer plots may not be permitted by your instructor. The idea is for you to learn how to plot data.

Remember always why graphs are plotted. Although it may seem quite a chore when first learning how to construct them, graphs are intended to improve your understanding of what you have determined in an experiment. Rather than just a list of numbers that may seem meaningless, a well-constructed graph can show you instantaneously whether your experimental data exhibit consistency and whether a relationship exists among them.

Vapor Pressure of Water at Various Temperatures

Temperature, °C	Pressure, mm Hg	Temperature, °C	Pressure, mm Hg
0	4.580	31	33.696
5	6.543	32	35.663
10	9.209	34	39.899
15	12.788	36	44.563
16	13.634	38	49.692
17	14.530	40	55.324
18	15.477	45	71.882
19	16.478	50	92.511
20	17.535	55	118.03
21	18.650	60	149.37
22	19.827	65	187.55
23	21.068	70	233.71
24	22.377	75	289.10
25	23.756	80	355.11
26	25.209	85	433.62
27	26.739	90	525.77
28	28.349	95	633.91
29	30.044	100	760.00
30	31.823		

APPENDIX E

Concentrated Acid–Base Reagent Data

Acids

Reagent	HCl	HNO₃	H₂SO₄	CH₃COOH
Formula weight of solute	36.46	63.01	98.08	60.05
Density of concentrated reagent	1.19	1.42	1.84	1.05
% concentration	37.2	70.4	96.0	99.8
Molarity	12.1	15.9	18.0	17.4
mL needed for 1 L of 1 M solution	82.5	63.0	55.5	57.5

Bases

Reagent	NaOH	NH₃
Formula weight of solute	40.0	35.06
Density of concentrated reagent	1.54	0.90
% concentration	50.5	56.6
Molarity	19.4	14.5
mL needed for 1 L of 1 M solution	51.5	69.0

Data on concentration, density, and amount required to prepare solution will differ slightly from batch to batch of concentrated reagent. Values given here are typical.

APPENDIX F

Density of Water at Various Temperatures

Temperature, °C	Density, g/mL	Temperature, °C	Density, g/mL
0	0.99987	26	0.99681
2	0.99997	28	0.99626
4	1.0000	30	0.99567
6	0.99997	32	0.99505
8	0.99988	34	0.99440
10	0.99973	36	0.99371
12	0.99952	38	0.99299
14	0.99927	40	0.99224
16	0.99897	45	0.99025
18	0.99862	50	0.98807
20	0.99823	55	0.98573
22	0.99780	60	0.98324
24	0.99732		

APPENDIX G

Solubility Infomation

Solubility Rules for Ionic Compounds in Water

- Nearly all compounds containing Na^+, K^+, and NH_4^+ are readily soluble in water.

- Nearly all compounds containing NO_3^- are readily soluble in water.

- Most compounds containing Cl^- are soluble in water, with the common exceptions of $AgCl$, $PbCl_2$, and Hg_2Cl_2.

- Most compounds containing SO_4^{2-} are soluble in water, with the common exceptions of $BaSO_4$, $PbSO_4$, and $CaSO_4$.

- Most compounds containing OH^- ion are not readily soluble in water, with the common exceptions of the hydroxides of the alkali metals (Group 1) and $Ba(OH)_2$.

- Most compounds containing other ions not mentioned above are not appreciably soluble in water.

APPENDIX H

Properties of Substances

Introduction

As a convenience to the user, some properties of the various chemical substances employed in this manual are listed in the table that follows. This table by no means lists all notable properties of the substances mentioned, and should it serve only as a quick reference to the substances. Additional information about the substances to be used in an experiment should be obtained before use of the substances. Sources of such additional information include the lab instructor or professor, the Material Safety Data Sheet (MSDS) for the substance, the label on the substance's bottle, and a handbook of dangerous or toxic substances.

In the following table, corrosiveness and toxicity risks are indicated as L (low), M (moderate), or H (high).

Substance	Flammability	Corrosiveness	Toxicity	Notes
acetaldehyde	Yes	M	M	1, 4
acetic acid	Yes	H	H	1, 4
acetic anhydride	Yes	H	H	
acetone	Yes	M	H	
acetylacetone	Yes	M	H	1, 4
adipoyl chloride	Yes	H	H	1, 4, 7
alanine	No	L	L	
aluminon reagent	No	M	M	
aluminum	Yes*	L	L	
ammonia	No	M	H	1, 4
ammonium carbonate	No	L	M	2
ammonium chloride	No	L	L	
ammonium molybdate	No	M	H	1
ammonium nitrate	No	M	H	3
ammonium sulfate	No	L	M	
ammonium sulfide	No	M	H	4
ammonium thiocyanate	No	L	H	
asparagine	Yes*	L	L	
barium chloride	No	M	H	
barium nitrate	No	M	H	3
Benedict's reagent	No	M	M	
benzoic acid, 2-amino-	Yes	M	H	
benzyl alcohol	Yes	L	M	
boric acid	No	L	M	
boron	Yes*	M	H	
bromine	No	H	H	1
butyl alcohol, n-	Yes	M	M	
butylamine, n-	Yes	M	M	1, 4
butyric acid	Yes	M	M	4
calcium	Yes*	M	L	7
calcium acetate	Yes	H	H	1, 4
calcium carbonate	No	L	L	2
calcium chloride	No	M	M	

Substance	Flammability	Corrosiveness	Toxicity	Notes
calcium hydroxide	No	M	M	
carbon	Yes*	L	M	
carbon disulfide	Yes	H	H	1, 4
chlorine	Yes*	H	H	1, 4
chromium(III) chloride	No	H	H	1
cobalt(II) chloride	No	M	H	
copper	Yes*	L	M	
copper(II) oxide	No	M	M	
copper(II) sulfate	No	M	M	
cyclohexane	Yes	L	H	
cyclohexene	Yes	M	H	4
dichlorobenzene, 1, 4-	Yes	M	H	
diethyl ether	Yes	M	H	
dimethylglyoxime	Yes	M	M	
dry ice	No	H	L	5
Eriochrome Black-T	No	M	H	
ethyl alcohol (ethanol)	Yes	L	H	6
ethyl ether	Yes	M	H	
ethylene diamine	Yes	M	H	1, 4
ethylene diamine tetraacetic acid (EDTA)	Yes*	M	H	
formaldehyde	Yes	M	H	1, 4
fructose	Yes*	L	L	
gelatin	Yes*	L	L	
glucose	Yes*	L	L	
glycine	Yes*	L	L	
graphite	Yes*	M	M	
hexane	Yes	M	H	
hexanediamine	Yes	H	H	4
hexene, 1-	Yes	M	M	
hydrochloric acid	No	H	H	1, 4
hydrogen	Yes	L	M	
hydrogen peroxide, 3%	No	H	M	3
iodine	No	H	H	
iron(II) sulfate	No	L	L	
iron(III) chloride	No	M	M	
iron(III) nitrate	No	M	M	3
isoamyl alcohol	Yes	L	M	
isobutyl alcohol	Yes	M	M	
isoleucine	Yes*	L	L	
isopropyl alcohol	Yes	L	M	
lead	Yes*	L	M	
lead acetate	Yes*	M	H	
lithium	Yes	H	H	7
lithium chloride	No	M	H	
magnesium	Yes	L	M	
magnesium chloride	No	L	M	
manganese(II) chloride	No	M	M	3
manganese(IV) oxide	No	M	M	3
mercury	No	H	H	
mercury(II) chloride	No	H	H	
methyl alcohol (methanol)	Yes	H	H	

Substance	Flammability	Corrosiveness	Toxicity	Notes
methyl salicylate	Yes	M	H	
methylene chloride	Yes	H	H	1, 4
n-butyl alcohol	Yes	M	M	
n-butylamine	Yes	M	M	4
n-octyl alcohol	Yes	M	M	
n-pentyl alcohol	Yes	M	M	
n-propyl alcohol	Yes	L	M	
naphthalene	Yes	M	H	
nickel(II) chloride	No	M	H	
nickel(II) sulfate	No	M	H	
ninhydrin	Yes	H	H	
nitric acid	No	H	H	1, 3, 4
nitrogen	No	L	L	
oleic acid	Yes	M	L	
oxygen	Supports	L	L	
pentane	Yes	M	H	
pentanedione, 2,4-	Yes	M	H	1, 4
pentyl alcohol, n-	Yes	M	M	
phenanthroline, 1,10-	Yes	M	M	
phosphoric acid	No	H	H	
phosphorus, red	Yes	H	H	
phosphorus, white	Yes	H	H	
potassium	Yes	H	H	7
potassium bromide	No	L	L	
potassium carbonate	No	L	L	
potassium chloride	No	L	L	
potassium chromate	No	H	H	3
potassium dichromate	No	H	H	3
potassium ferrocyanide	No	M	H	
potassium hydrogen phthalate	Yes*	M	M	4
potassium hydroxide	No	H	H	
potassium iodate	No	H	H	3
potassium iodide	No	M	M	
potassium nitrate	No	M	H	3
potassium oxalate	Yes*	M	H	
potassium permanganate	No	H	H	3
potassium sulfate	No	L	L	
proline	Yes*	L	L	
propionic acid	Yes	H	H	
propyl alcohol, n-	Yes	L	M	
salicylic acid	Yes*	M	M	
serine	Yes*	L	L	
silicon	No	M	M	
silver acetate	Yes*	M	M	
silver chromate	No	H	H	3
silver nitrate	No	M	M	
sodium	Yes	H	H	7
sodium acetate	Yes*	M	M	
sodium bicarbonate	No	L	L	
sodium bromide	No	L	L	
sodium chloride	No	L	L	
sodium chromate	No	H	H	3

Substance	Flammability	Corrosiveness	Toxicity	Notes
sodium hydrogen sulfite	No	M	M	4
sodium hydroxide	No	H	H	
sodium hypochlorite	No	H	H	1, 4
sodium iodide	No	M	M	
sodium nitrate	No	H	H	3
sodium nitrite	No	H	H	3
sodium nitroprusside	No	M	M	
sodium silicate	No	L	L	
sodium sulfate	No	L	L	
sodium sulfite	No	M	M	3, 4
sodium thiosulfate	No	M	M	
starch	Yes*	L	L	
strontium chloride	No	M	M	
sulfur	Yes	L	L	
sulfuric acid	No	H	H	3, 7
tartaric acid	Yes*	M	L	
thioacetamide	Yes*	H	H	4
tin	No	L	L	
tin(II) chloride	No	M	M	3
toluene	Yes	H	H	
zinc	Yes*	L	L	
zinc sulfate	No	L	M	

*Will burn if deliberately ignited, but has very low vapor pressure and is not a fire hazard under normal laboratory conditions. Such substances are more precisely termed "combustible" rather than flammable.

Notes:

1. This substance is a respiratory irritant. Confine use of the substance to the exhaust hood.

2. This substance vigorously evolves carbon dioxide if acidified.

3. This substance is a strong oxidant or reductant. Although the substance itself may not be flammable, contact with an easily oxidized or reduced material may cause a fire or explosion. Avoid especially contact with any organic substance.

4. This substance possesses or may give rise to a stench. Confine use of this substance to the exhaust hood.

5. Handle dry ice with tongs to avoid frostbite.

6. Ethyl alcohol for use in the laboratory has been denatured. It is not fit to drink and cannot be made fit to drink.

7. This substance reacts vigorously with water and may explode.